MW01147130

Medical Imaging Technology

Medical Imaging Technology

Victor I. Mikla and Victor V. Mikla

Chair of Physical and Mathematical Disciplines, Humanitarian and Natural Sciences, University of Uzhgorod

AMSTERDAM • BOSTON • HEIDELBERG • LONDON • NEW YORK • OXFORD
PARIS • SAN DIEGO • SAN FRANCISCO • SINGAPORE • SYDNEY • TOKYO

Elsevier
32 Jamestown Road, London NW1 7BY
225 Wyman Street, Waltham, MA 02451, USA

British Library Cataloguing-in-Publication Data
A catalogue record for this book is available from the British Library

Library of Congress Cataloging-in-Publication Data
A catalog record for this book is available from the Library of Congress

ISBN: 978-0-12-417021-6

For information on all Elsevier publications
visit our website at store.elsevier.com

This book has been manufactured using Print On Demand technology. Each copy is produced to order and is limited to black ink. The online version of this book will show color figures where appropriate.

Epigraph

The first gulp from the glass of natural sciences will turn you into an atheist, but at the bottom of the glass **God** is waiting for you.

Werner Heisenberg, Nobel Laureate in Physics (as cited in Hildebrand, 1988, 10)

Contents

Preface

Can physics be interesting or exciting for physicians, medicals?

Personally, I find that most textbooks on physics are dry, confusing, and serve only as a useful cure for my insomnia.

This book is different. I know (at least I hope), what works, what does not, and how to present information clearly.

A particularly strong point of this book—it covers all areas of medical diagnostic imaging.

Surely, medical imagers are more at home with pictures rather than text and formulas. Most authors of other physics books have not grasped these concepts.

During the last few decades of twentieth century, partly due to the increasing availability of relatively inexpensive computational resources, medical imaging technology which had for nearly 90 years been almost exclusively concerned with film/screen X-ray imaging experienced the development and commercialization of new imaging technologies. Computed tomography, magnetic resonance imaging, digital subtraction angiography, Doppler ultrasound imaging, and various imaging technologies based on nuclear emission (PET, SPET, etc.) have all been valuably additions to the radiologist's arsenal of imaging tools toward even more reliable detection and diagnosis of disease.

As evident from research described in this book, medical imaging is still undergoing very rapid change. The authors hope that this publication will provide at least some of physics information required by students, researchers, lecturers, and practioners in this exciting field to make their own contributions to its ever-increasing usefulness.

Victor I. Mikla, Prof., PhD, DSc
Chair of Physical and Mathematical Disciplines
Uzhgorod National University
Uzhgorod
4th May, 2013

Acknowledgment

A number of people helped in a significant way to make this book a reality. First and foremost, we would like to give our special thanks to Dr. Lisa Tickner (Publishing Director at Elsevier), for insisting on this book, for her confidence, and for continual support throughout the writing process. We would like to thank Dr. Erin Hill-Parks, who also applied her very expert skill as an editor and to Sarah Lay as a Project Manager.

1 Advances in Imaging from the First X-Ray Images

1.1 X-Rays—Early History

The true start of imaging, medical and nonmedical, as we consider it today was in 1895 when Roentgen discovered the X-ray in his Wurzburg laboratory on November 8. Almost immediately after Wilhelm Conrad Roentgen announced the discovery of the X-ray, imaging techniques based on the discovery were implemented all over the world. For example, chest radiography became one of those early applications, even though the equipment seems crude by comparison to that of today, and certainly there was no knowledge of the potential deleterious effects of the ionizing radiation of X-rays. Roentgen's invention heralded the start of the modern era of medical imaging.

German physicist, a person who studies the relationship between matter and energy, Wilhelm Conrad Roentgen (1845–1923) discovered X-rays in 1895 while he was experimenting with electricity (Figure 1.1). Because he did not really understand what these rays were, he called them X-rays. By 1900, however, doctors were using X-rays to take pictures (called radiographs) of bones, which helped them treat injuries more effectively. In 1901, Roentgen was awarded the first Nobel Prize for physics for his discovery of this short-wave ray. He donated all the prize money to his university. Keeping his will the newly discovered rays were called just X-rays ("x" is usually used to denote the unknown entity). He refused to take out patent on this discovery because he wanted the entire humankind to receive the benefits of X-rays. He even forbade naming of these rays after him. The reader can find historical details that describe events leading up to and occurring around the time of Roentgen's discovery in excellent articles by Linton [1], Assmus [2], and Mould [3], to say about few. Roentgen tested his new rays to define their properties and made numerous observations before writing a paper that was submitted for publication on December 28, 1895, to a local scientific publication [4]. Roentgens prestige in the community was such that his paper was accepted without review and published immediately [4].

Scientific American published in its new section a short bulletin headlined "Professor Roentgen's Wonderful Discovery" [5]. The first mention of the X-ray in *Science* magazine was a brief latter in the issue of January 31, 1896, from Hugo Munsterberg, a Harvard University physics professor [6]. *Nature* published a full translation of Roentgens paper [7]. Roentgen was dominated by a shyness that

Medical Imaging Technology. DOI: http://dx.doi.org/10.1016/B978-0-12-417021-6.00001-0

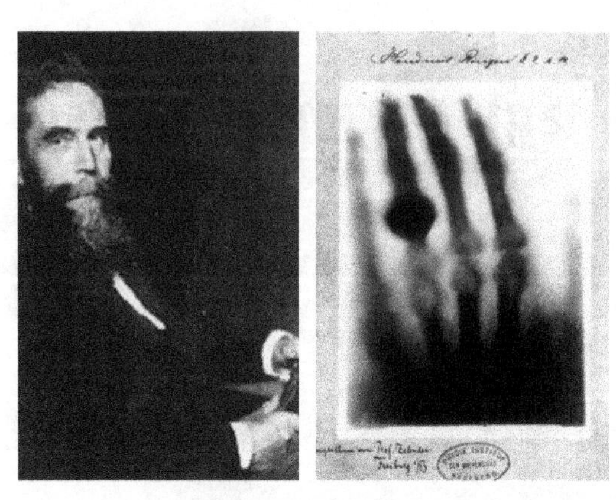

Figure 1.1 Wilhelm Conrad Roentgen (at the left) and Roentgen's by-then-famous first X-ray image of his wife, Anna Bertha, hand (at the right). The X-rays penetrated through Anna's skin, muscles, and other tissues but could not go through bones and wedding ring in her finger.

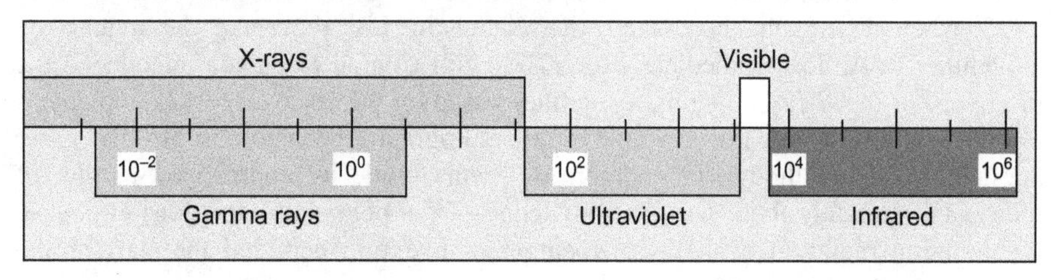

Figure 1.2 X-rays are part of the electromagnetic spectrum (wavelength is given in angstroms). X-rays have a wavelength in the range of 0.01–10 nm, corresponding to frequencies in the range of 3×10^{16}–3×10^{19} Hz and energies in the range of 100 eV to 100 keV. They are shorter in wavelength than UV rays and longer than gamma rays.

made him evade personal contacts wherever he could. In a brief period of time that follows, the X-ray went from a physics experiment to almost routine medical practice.

One can see what could be achieved in physics before the occurrence of the technological revolution involving not only computer applications but also the disappearance of the small independent X-ray companies into today's multinational companies. Research and development is nowadays just too expensive for much independent practical high-technology contributions without financial backing.

1.2 The Use of X-Rays for Analysis

X-rays are a form of electromagnetic radiation, as is light (Figure 1.2). Their distinguishing feature is their extremely short wavelength—only about 1/10 000 that of light (or even less). This characteristic is responsible for the ability of X-rays to penetrate materials that usually absorb or reflect ordinary (visible) light.

X-rays exhibit all the properties inherent of light, but in such a different degree as to modify greatly their practical behavior. For example, light is refracted by glass and, consequently, is capable of being focused by a lens in such instruments as cameras, microscopes, telescopes, and spectacles. X-rays are also refracted, but to such a very slight degree that the most refined experiments are required to detect this phenomenon. Hence, it is impractical to focus X-rays. It would be possible to illustrate the other similarities between X-rays and light but, for the most part, the effects produced are so different—especially their penetration—that it seemed preferable to consider X-rays separately from other radiations. Figure 1.2 shows their location in the electromagnetic spectrum.

Their similarities to light led to the tests of established wave optics: polarization, diffraction, reflection, and refraction. Despite limited experimental facilities Roentgen could find no evidence of these. May be, it was additional reason he called them "x" (unknown) rays.

For diffraction applications, only short wavelength X-rays (hard X-rays) in the range of a few Angstroms to 0.1 Å (1 − 120 keV) are used. Because the wavelength of X-rays is comparable to the size of atoms, they are ideally suited for probing the structural arrangement of atoms and molecules in a wide range of materials. The energetic X-rays can penetrate deep into the materials and provide information about the bulk structure.

The year 2012 marks the 100th anniversary of the discovery of X-ray diffraction (XRD) and its use as a probe of the structure of matter [8−11]. Sixteen years after Rontgen announced in 1895 his discovery of "X" rays that can penetrate the body and photograph its bones, Max von Laue, a professor of physics at the University of Munich in Germany, worked on a theory of the interference of light in plane parallel plates. Laue was Plank's favorite disciple. His interest covers the whole of physics. If, as some argued, X-rays were not made up of particles but were a form of electromagnetic radiation similar to ordinary (visible) light, and then it should be possible to repeat well-known optical experiments using X-rays instead of beams of ordinary light.

In 1911, Laue suggested to one of his research assistants, Walter Friedrich, and a doctoral student, Paul Knipping, that they try out X-rays on crystals. His reasoning was that X-rays have a wavelength close to the interatomic distances in crystals, and as a result, the crystal should act as a diffraction grating (Figure 1.3). Laue, always the theoretician, did not actually make the necessary experiments. In a single elegant experiment performed by Friedrich and Knipping, Laue had proven the wave-like nature of X-rays and the space−lattice structure of crystals at the same time [8,9]. When he received the Nobel Prize for what the Committee said was his "epoch-making discovery", Laue gratefully acknowledged Friedrich and Knipping for their roles in the discovery; and for him it went without saying that he shared his prize money with them [10,11]. Einstein hailed Laue's discovery as one of the most beautiful in the history of physics.

Before the Laue experiment, did anyone dream of tools allowing one to explore the structure of matter on the molecular scale and use this information for deriving structure−property relationships of materials—or for understanding the molecular basis of life? After X-rays had already been used to image the internal anatomy of the human body, with the Laue experiment the internal structure of crystals also became accessible on nanoscale [11].

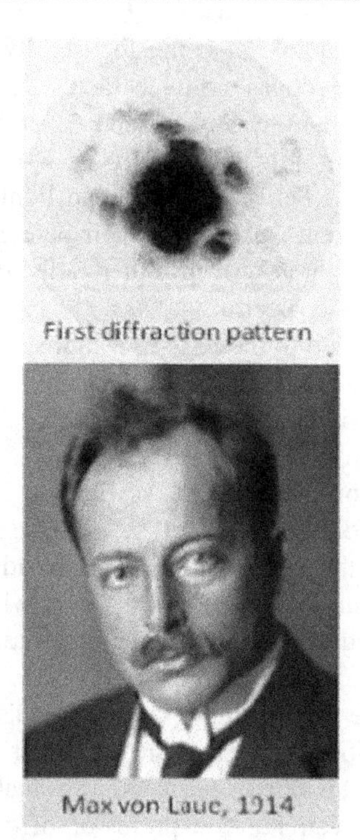

Figure 1.3 Max von Laue and the first diffraction pattern.

Laue's pioneering work in X-ray crystallography opened the way for two, quite different, developments in physics, both of them of immense importance. First, it confirmed the electromagnetic nature of X-radiation and made it possible to determine the wave length of X-rays with great accuracy. Second, it gave physicists and chemists a new tool for investigating the atomic structure of matter. In the 1950s, it was XRD studies that enabled scientists to reveal the structure of the nucleic acids (DNA and RNA) and to establish the new discipline of molecular biology.

English physicists Sir William Henry Bragg and his son, Sir William Lawrence Bragg, argued that when the X-rays are reflected off two successive planes of atoms in the crystal, they interfere constructively if the difference in the distance traveled is equal to an integral number of wavelengths. Thus, the famous Bragg condition is

$$n\lambda = 2d \sin \theta$$

They have developed the above relationship to explain why the cleavage faces of crystals appear to reflect X-ray beams at certain angles of incidence (θ) [12]. The variable d is the distance between atomic layers in a crystal, and the variable λ is the wavelength of the incident X-ray beam; n is an integer. This observation is an example of X-ray wave interference, commonly known as XRD, and was direct evidence for the periodic atomic structure of crystals. By 1913, just a year after they had pioneered the method, crystal analysis with X-rays had become a standard method.

Figure 1.4 Sir William Henry Bragg and Sir William Lawrence Bragg.

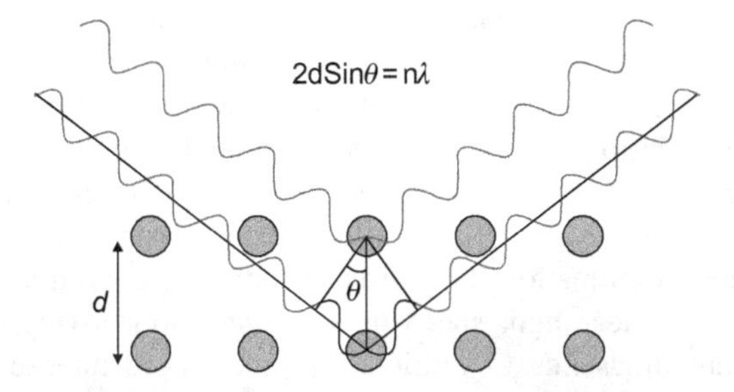

Figure 1.5 Bragg's law: 2D representation. An incident X-ray beam interacts with the atoms arranged in a periodic manner. The atoms (green spheres) can be viewed as forming different sets of planes in the crystal (colored lines). For a given set of lattice planes with an interplane distance d, the condition for a diffraction to occur can be simply written $2d \sin \theta = n\lambda$. In the equation, λ is the wavelength of the X-ray, θ is the scattering angle and n is the order of interference. (For interpretation of the references to color in this figure legend, the reader is referred to the web version of this book.)

The results gave insight into the structure of crystals. The Braggs were awarded the Nobel Prize in physics in 1915 for their work in determining crystal structures beginning with NaCl, ZnS, and diamond (Figure 1.4) [2,11–13].

Bragg's law describes the mechanism by which XRD occurs and was an extremely important discovery—it formed the basis for what is now known as crystallography (Figure 1.5).

X-rays primarily interact with electrons in atoms. When X-ray photons collide with electrons, some photons from the incident beam will be deflected away from the direction where they originally travel. If the wavelength of these scattered X-rays did not change (this means that X-ray photons did not lose any energy), the process is called elastic scattering in that only momentum has been transferred in the scattering process. These are the X-rays that one measure in diffraction experiments, as the scattered X-rays carry information about the electron distribution in materials. On the other hand, in the inelastic scattering process, X-rays transfer some of their energy to the electrons and the scattered X-rays will have different wavelength than the incident X-rays.

Diffracted waves from different atoms can interfere with each other and the resultant intensity distribution is strongly modulated by this interaction. If the atoms are arranged in a periodic fashion, as in crystals, the diffracted waves will consist of sharp interference peaks with the same symmetry as in the distribution of atoms. Measuring the diffraction pattern therefore allows us to deduce the distribution of atoms in a material. The peaks in an XRD pattern are directly related to the atomic distances.

In the same year, Moseley showed the wavelengths were not only characteristic of the element the target was made of, but also they had the same sequence as the atomic numbers. This allowed atomic numbers to be determined unambiguously for the first time.

X-ray crystallography is a standard technique for solving crystal structures. Its basic theory was developed soon after X-rays were first discovered more than a century ago. Over the years that follow it has gone continual development in data collection, instrumentation, and data reduction methods. In recent years, the advent of synchrotron radiation sources, area detector-based data collection instruments, and high-speed computers has dramatically enhanced the efficiency of crystallographic structural determination. Synchrotron radiation is emitted by electrons or positrons traveling at near light speed in a circular storage ring. These powerful sources, which are thousands to millions of times more intense than laboratory X-ray tubes, have become indispensable tools for a wide range of structural investigations and brought advances in numerous fields of science and technology [14].

Today X-ray crystallography is widely used in materials and biological research.

In X-ray crystallography, integrated intensities of the diffraction peaks are used to reconstruct the electron density map within the unit cell in the crystal. To achieve high accuracy in the reconstruction, which is done by Fourier transforming the diffraction intensities with appropriate phase assignment, a high degree of completeness as well as redundancy in diffraction data is necessary, meaning that all possible reflections are measured multiple times to reduce systematic and statistical error. The most efficient way to do this is by using an area detector. The latter can collect diffraction data in a large solid angle.

The most common use of powder (polycrystalline) diffraction is chemical analysis. This can include phase identification, investigation of high (low) temperature phases, solid solutions, and determinations of unit cell parameters of new materials. The crystalline inclusions in inorganic and organic polymers give sharp narrow diffraction peaks and the amorphous component gives a very broad peak (halo). The ratio between these intensities can be used to calculate the amount of crystallinity in the material.

Nowadays, XRD allows a range of determinations to be made including phase identification of crystalline materials, phase quantification, glass content, and quality control methods, to say about few.

Soon after it was also established that secondary fluorescent X-rays were excited in any material irradiated with beams of primary X-rays. This started investigation into the possibilities of fluorescent X-ray spectroscopy as a means of qualitative and quantitative elemental analysis.

1.3 Industrial Radiology

X-rays are also used extensively in industry as a nondestructive testing method that examines the volume of a specimen. Radiographs of a specimen showed any changes in thickness, internal and external defects, and assembly details invisible to the naked eye.

The use of ionizing radiation, particularly in medicine and industry, is growing throughout the world, with further expansion likely as technical developments result from research. One of the longest established applications of ionizing radiation is industrial radiography, which uses both X-radiation and gamma radiation to investigate the integrity of equipment and structures. Industrial radiography is widespread in almost all developed countries. It is indispensable to the quality assurance required in modern engineering practice and features in the work of multinational companies and small businesses alike.

To verify the quality of a product, samples are taken for examination or a nondestructive test (NDT) is carried out. In particular, with fabricated (welded) assemblies, where a high degree of constructional skill is needed, it is necessary that nondestructive testing is carried out.

Most NDT systems are designed to reveal defects, after which a decision is made as to whether the defect is significant from the point of view of operational safety and/or reliability. Acceptance criteria for weld defects in new constructions have been specified in standards. However, NDT is also used for purposes such as the checking of assembled parts, the development of manufacturing processes, the detection of corrosion, or other forms of deterioration during maintenance inspections of process installations and in research.

Industrial radiography is extremely versatile. The equipment required is relatively inexpensive and simple to operate. It may be highly portable and capable of being operated by a single worker in a wide range of different conditions, such as at remote construction sites, offshore locations, and cross-country pipelines as well as in complex fabrication facilities. The associated hazards demand that safe working practices be developed in order to minimize the potential exposure of radiographers and any other persons who may be in the vicinity of the work. The use of shielded enclosures (fixed facilities), with effective safety devices, significantly reduces any radiation exposures arising from the work.

The demands and rewards of industrial radiography, the ready availability of the essential equipment, the wide range of working conditions, and the fact that the techniques employed usually involve the routine manipulation and exposure of powerful gamma emitting sources and X-ray machines have all been identified as contributory to the likelihood of accidents. Even in the United States of America and EU-Member States with highly developed regulatory infrastructures, industrial radiographers, on average, receive radiation doses that exceed those of other occupationally exposed workers, and individual industrial radiographers are the most likely group of workers to receive doses approaching relevant dose limits. Radiation protection and safety in industrial radiography is thus of great importance in both developed and developing countries.

Presently, a wide range of industrial radiographic equipment, image forming techniques, and examination methods are available. Skill and experience are needed to select the most appropriate method for a particular application.

The ultimate choice will be based on various factors such as the location of the object to be examined, the size of the NDT equipment, the image quality required, the time available for inspection, and last but not least financial considerations [15,16].

1.4 Airport Security

In the 1960s, X-ray screening machines were introduced alongside metal detectors at airports to detect bombs in luggage. Since then, they have become a standard fixture not only in airports but also in many government buildings.

Security scanners are a valuable alternative to existing screening methods and an effective method of screening passengers as they are capable of detecting both metallic and nonmetallic items carried on a person. The scanner technology is developing rapidly and has the potential to significantly reduce the need for manual searches applied to passengers, crews, and airport staff.

In order not to risk jeopardizing citizens' health and safety, only security scanners which do not use X-ray technology are added recently to the list of authorized methods for passengers screening at EU airports.

1.5 The Use of X-Rays for Medical Purposes

1.5.1 Retrospective Glance

The medical community immediately recognized and continuously developed the extraordinary potential of X-rays for diagnostic purposes. Physicians reading about seeing bones on X-rays were quick to perceive medical applications. Edwin Frost's physician brother brought him the patient with a fractured ulna who received the first documented USA medical X-ray [1]. In the decades following, several applications demonstrated for the first time the ability to look inside the body without dissection to study internal anatomy. Over the years there have been significant improvements in the X-ray technique, primarily due to the higher sensitivity and fidelity of the recording films, not necessarily any significant advances in the basic methodology.

This section attempts to briefly summarize the history of visualizing the internal body for medical, with primary focus on present capabilities. A few predictions will be made by extrapolating from present to possible future advances. Copious citations will not be used, as most of this treatise is based on four decades of personal opinion and experience as well as the topics included in this perspective—past, present, and future.

If we focus on the historical evolution of medical imaging alone, leaving aside for now the significant parallel advances in biological imaging facilitated by the invention of the microscope, the field dates back to the early parts of the twelfth and thirteenth centuries with direct visualization by dissection in anatomy theaters. This was the principal form of imaging that is direct visualization via dissection, for almost 600 years until near the end of the nineteenth century when a form of imaging was introduced to aid visualization into the body without dissection—namely the discovery of the X-ray. The sensitivity and quality of recordings by this technique improved over the next several decades.

For many years after the discovery of X-rays, X-ray fluoroscopic examinations were performed in a dark room by observing images on a fluorescent screen that glowed in response to X-rays that had passed through the patient's body.

Roentgen's revolutionary discovery gave birth to the professions of diagnostic and therapeutic medical physics. Since that time, physicists have worked avidly to develop new discoveries to advance the technology of medical imaging and radiation therapy. The first 70 years after Roentgen's discovery witnessed the development of higher speed imaging systems, electronic amplification devices, scintillation cameras, ultrasonographic (US) devices, advanced high capacity X-ray tubes, and rapid film processors. However, maturation of the computer has accelerated even more and enabled technology such as computed tomography (CT), magnetic resonance imaging (MRI), and sophisticated interventional fluoroscopy.

In the early days of radiology, equipment was quite primitive with little or no shielding around the X-ray tube and bare metal high-voltage cables strung across the ceiling (Figure 1.6). Often the physician responsible for the "X-ray laboratory"

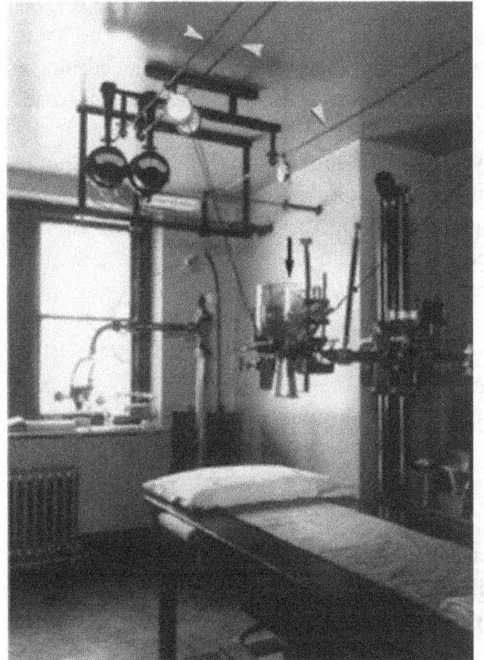

Figure 1.6 An X-ray examination room (Mayo Clinic, Rochester, circa 1925) with bare high-voltage cables (arrowheads) and little shielding of the X-ray tube (arrow) [17].

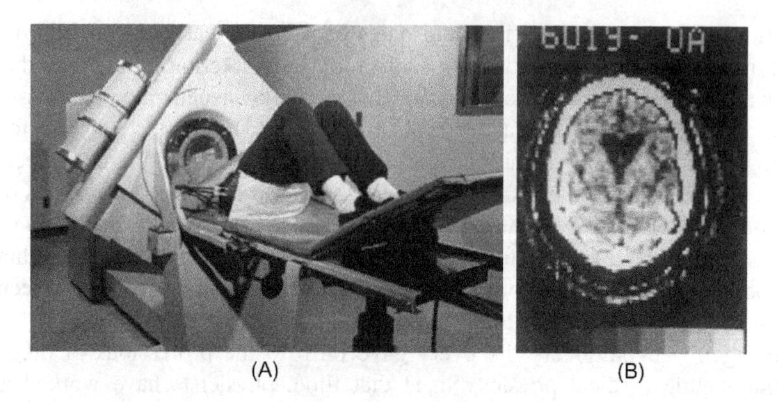

Figure 1.7 (A) Original EMI CT head scanner and (B) an 80 × 80-matrix head CT image obtained with it [17].

served as the technologist, service engineer, and medical physicist to ensure that the equipment was functional.

Chronologically, a major step forward in fluoroscopy was made by replacing the fluorescent screen with an electronic instrument based on television technology in the 1950s. This leads to a significant increase in image brightness.

Major changes in the practice of medical physics started in the early 1970s with the introduction of the CT scanner. Gray recalls being in the equipment exhibit hall shortly after the exhibits opened at the meeting of the Radiological Society of North America (RSNA) at the Palmer House Hotel (Chicago) in 1972. The original EMI CT scanner was on display and the author [17] quickly judged it to be something that the radiology community would not embrace because it produced images with checkerboard-sized pixels (Figure 1.7). The author [17] wondered why a radiologist would be interested in something that produced images with such poor spatial resolution. Since then, medical physicists have learned that other image quality parameters, in addition to spatial resolution, are important in diagnostic imaging.

In the 1980s, digital subtraction angiography (DSA) was introduced in vascular contrast studies. In DSA, analog video signals from a TV camera are converted to digital data, and vascular structures are clearly demonstrated through digital processing and subtraction of nonvascular anatomy. Subsequently, rapid advances in computer technology resulted in a wider range of diagnostic applications. Many well-known medical physicists practicing in the twentieth century made notable contributions to diagnostic and therapeutic medical physics (Tables 1.1 and 1.2).

1.5.2 Radiography/Fluoroscopy

Radiological imaging is a process by which the attenuation of an X-ray beam traversing a part of a human body is either recorded in a medium for later medical interpretation of potential pathology (**radiography**) or displayed in real time on a monitor for functional assessment (**diagnostic fluoroscopy**) or intervention

Table 1.1 A Few of the Technologic Advances in Diagnostic Imaging in the Past Century [17]

Year	Technological Advance	Prominent Pioneers
1895	Discovery of X-rays	W. C. Roentgen
1896	Calcium tungstate screens	T. A. Edison
1896	Discovery of radioactivity	H. Becquerel
1913	Hot-cathode X-ray tube	W. D. Coolidge
1915	Bucky Potter grid	G. P. Bucky, H. E. Potter
1925	Dual-emulsion film on flexible base	
1928	Pako mechanized film processor	
1934	Conventional tomography	A. Vallebona, G. Z. DesPlantes
1948	Westinghouse image intensifier	J. W. Coltman
1951	Rectilinear scanner	B. Cassen
1951	Bistable US	G. Ludwig, J. Wild, D. Howry
1956	Kodak X-Omat film processor	
1958	Scintillation camera	H. Anger
1960	Xeroradiography	
1962	Emission reconstruction tomography	D. Kuhl
1962	Gray-scale US	G. Kosoff
1965	Dedicated mammographic system	
1972	CT	G. N. Hounsfield
1973	Digital subtraction angiography	C. Mistretta
1975	Positron emission tomography or PET	M. Ter-Pogossian
1976	Single photon emission CT or SPECT	J. Keyes
1980	MR imaging	P. C. Lauterbur

(**interventional radiology**). New detectors of the computed radiography (CR) or the digital radiology (DR) type are steadily replacing screen/film (S/F) combinations in radiography and image intensifiers and video cameras in fluoroscopy.

As the primary X-ray absorber, S/F radiography uses luminescent intensifying screens, while CR uses storage phosphor screens. In S/F, the emitted light is the only image signal available to the film, the recording medium. Image processing in DR is different. There are two types of DR systems:

1. those that are based on charge-coupled devices (CCDs);
2. those that are based on amorphous silicon (a-Si:H) solid-state flat-panel detectors.

Direct DR technologies convert the incident X-ray quanta into a measurable latent image signal, while indirect DR technologies require some intermediate steps. A direct detector produces charge directly from the X-ray absorption within a semiconductor (e.g., a-Se).

An indirect detector produces secondary signals, such as light photons from an X-ray scintillator, for subsequent conversion to charge by a thin-film transistor (TFT) photodiode array or CCD camera. Flat-panel detectors are transforming both radiographic and fluoroscopic imaging and are an integral component of image-guided radiotherapy.

Table 1.2 A Few of the Technologic Advances in Radiation Therapy in the Past Century [17]

Year	Technologic Advances	Prominent Pioneers
1895	Discovery of X-rays	W. C. Roentgen
1896	Discovery of radioactivity	H. Becquerel
1898	Discovery of radium	M. and P. Curie
1913	Development of hot-cathode X-ray tube	W. D. Coolidge
1914–1917	First radon plants established	W. Duane, G. Failla
1921	First developments in radium dosimetry	R. Sievert, E. Quimby
1928	Establishment of the roentgen as unit of "dose"	
1933	First treatments with Van de Graaf generator	J. Trump
1943–1948	First betatron treatments	D. Kerst, G. Adams, J. Laughlin, H. Johns
1951	First cobalt 60 treatments	H. Johns, L. Grimmett
1953	First linear accelerator treatments	M. Day, F. Farmer
1958	Computerized treatment planning introduced	J. Laughlin, T. D. Sterling, K. C. Tsien, R. Wood
1959	First remote afterloading units	B. Proimos, K. Wright, J. Trump, W. Jennings, T. Davy, J. Brace, A. Green
1960	First remote afterloading units	R. Walstam, U. Henschke
1962	Electronic portal imaging introduced	S. Benner
1964	High-dose-rate remote afterloaders introduced	U. Henschke, R. Walstam
1965	Conformal radiation therapy with multileaf collimation introduced	S. Takahashi
1968	Gamma knife introduced	*Leksell
1968	Radiological physics center established†	R. Shalek
1969	First commercial treatment planning systems	R. Bentley, J. Cox, W. Powers
1980	First "modern" electronic portal imager	N. A. Baily
1984	First "modern" multileaf collimator	A. Brahme, J. Mantel, H. Perry

The exposure required to form an image varies depending on the sensitivity of the image receptor as shown in Figure 1.8.

Other characteristics of the image are contrast and contrast sensitivity (the ability to visualize low-contrast objects), blurring and visibility of detail (spatial resolution), visual noise, artifacts, and spatial (geometric) characteristics (magnification and distortion). The major controlling factor for X-ray image noise is the amount of

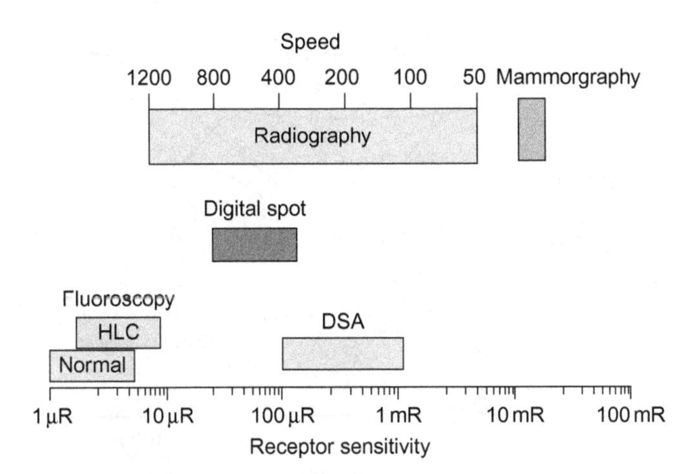

Figure 1.8 X-ray receptor sensitivity. Receptors for X-ray imaging cover a wide range of sensitivity or speed values [18]. This has an effect on the noise in the images captured by each (DSA = digital subtraction angiography, HLC = high level control operating mode (1 mR–10 μGy)).
Source: From [18].

exposure to the receptor. With screen–film radiography, the receptor sensitivity is fixed to a specific value determined by the type of film and intensifying screens used and the quality of the film processing. Because of the relatively narrow film latitude, the exposure to the receptor must be within a limited range, or the films will be either under- or overexposed and appear too light or too dark. Therefore, the noise level in a film radiograph is determined by the design characteristics of the film and the screens that give it a specific sensitivity (speed). On the other hand, digital radiographic receptors have a wide dynamic exposure range (Figure 1.9).

Since its inception, fluoroscopy has and continues to be a key imaging modality used in interventional radiology, a branch of radiology that is concerned with the use of image guidance to conduct minimally invasive procedures for both diagnostic and therapeutic purposes. These procedures include angiography, angioplasty, pacemaker insertion, and embolization, to say about few.

The reason X-ray fluoroscopy remains a dominant imaging modality in interventional radiology is because no other single modality provides the same combination of high spatial and temporal resolution which is particularly important for proper deployment of endovascular (from within the blood vessel) devices such as stents or coils.

1.5.3 Clinical Fluoroscopy Requirements for Interventional Radiology

A modern fluoroscope consists of a large "C" shaped mount called a C-arm with an X-ray source on one end and an X-ray imager on the other. This assembly can be positioned such that different projections of the patient anatomy may be acquired from different angles. In certain cases, the C-arm assembly is rotated

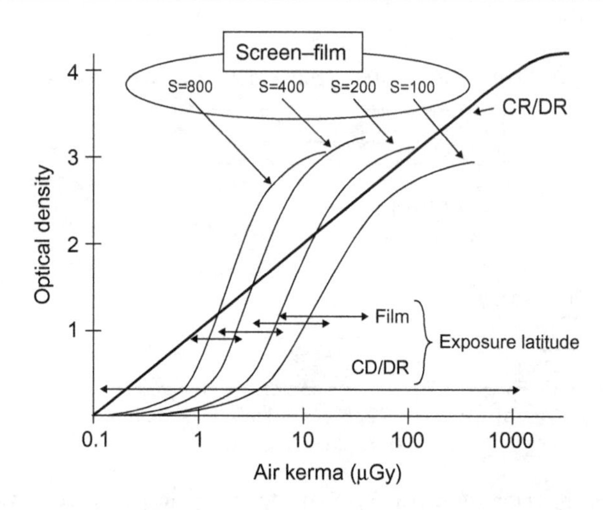

Figure 1.9 Response of the film to variations in radiation exposure is nonlinear [18]. Contrast enhancement and radiographic speed are based on the characteristics of the intensifying screen (phosphor) and the design of the light-sensitive silver halide emulsion to respond to chemical processing. A selection of screen–film speeds can be chosen to achieve appropriate optical density for a given dose (SNR). CR and DR exhibit a linear response over 10,000 in exposure latitude to allow the user to achieve an SNR sufficient to enable a reliable diagnosis.

around the patient during injection of a contrast dye and multiple projection images are acquired and subsequently reconstructed into a three-dimensional (3D) rendition of the vasculature. The C-arm assembly should hence be designed such that the X-ray source and X-ray imager are as small and light as possible to facilitate its positioning or rotation and to improve patient accessibility.

In the fluoroscopic mode, the physician typically guides interventional devices such as guide wires or stents through a catheter toward the lesion and is mainly concerned with tracking the position of the device. This guidance, usually referred to as a ***catheterization***, is typically done at relatively high imaging frame rates (7.5–30 image frames per second) so that the physician can get a good sense of the advancement of the device toward the lesion while avoiding potential complications due to, for example, arterial tortuosity. Catheterization is often a relatively time-consuming process (10 min or longer) and is done at very low X-ray exposures (mean exposure of 1 μR per frame at the imager) to minimize patient radiation exposure. The X-ray exposures employed are so low, in fact, that quantum noise (the stochastic variation in the spatial distribution of X-ray photons) becomes the dominant form of noise in the fluoroscopic images obtained using an X-ray image intensifier (XRII) system. Because the interventional device being imaged typically has a high degree of radioopacity, these low exposures are often adequate to obtain sufficient contrast to image the device. In the cine mode, sequences of images are acquired during administration of an X-ray contrast agent (typically via a catheter) into the vasculature. These relatively short (several seconds in duration) image acquisitions are taken at higher exposures (~10 μR/frame) so as to provide superior image quality (less quantum noise) and thus improve the diagnostic value of the images. This mode is also used when accurate positioning or deployment of endovascular devices is performed. In the radiographic mode, images are acquired at even larger exposures (~100 μR per frame) for applications such as DSA which have been shown to improve the detection of certain lesions such as aneurysms or thrombi.

High spatial resolution is an important requirement in the cine and radiographic imaging modes, as it can strongly affect the diagnosis or treatment outcome of an interventional procedure. Interventional devices such as guide wires or stenos typically have wire diameters ranging from 50 to 200 μm. It has been demonstrated that despite the effects of X-ray scattering, imaging of individual stenos wires (struts) using a high-resolution imager is possible inside a human head phantom; this should enable the deployment of novel asymmetrical stenos for specialized therapeutic neurovascular applications. Furthermore, in certain applications such as coronary angiography, detection of small calcium deposits (tens of micrometers in size) in coronary arteries provides an important means of assessing the degree of atherosclerosis as well as the likelihood of a successful angioplasty. For optimal imaging of fine features in interventional radiology, the imager should be able to resolve 5 line pairs per millimeter (a line pair is a pair of light and dark lines) such that the relative contrast between the two lines of each pair is greater than 0.2 (i.e., a modulation transfer function (MTF) greater than 0.2 at a spatial frequency of 5 cycles/mm).

From the earlier discussion, it follows that the key requirements for a clinical X-ray imager for interventional radiology—beyond a thin profile and providing unobstructed access to the patient—are:

1. quantum-noise limited (QNL) operation at the lowest clinical fluoroscopic X-ray exposures (in conformance with the ALARA principle);
2. capability for modes of operation that require significantly higher X-ray exposures; and
3. capability of imaging fine features of interventional devices or lesions.

The imager should also be able to operate at up to 30 frames per second.

1.5.4 X-Ray Image Intensifiers

Until recently, the most widely used X-ray imaging system in interventional radiology was the XRII. This is an electrooptical device that operates inside a vacuum enclosure (Figure 1.10A) which contains an input phosphor used to convert X-rays into optical photons.

The phosphor is coupled to a photocathode which, upon exposure to these optical photons, produces electrons. The latter are accelerated in an electric field and hit an output phosphor screen. This process enables the production of several thousand optical photons for each photoelectron emitted from the photocathode. The resulting optical image is captured using an optical assembly and a video camera or CCD. XRII/video systems provide excellent *X-ray sensitivity* (the degree to

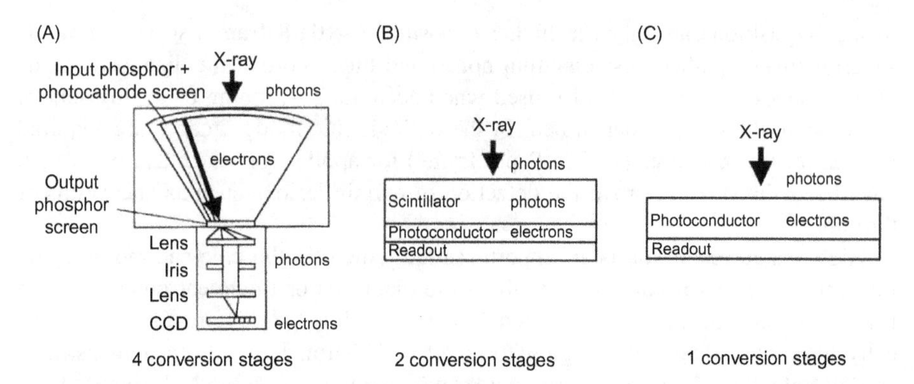

Figure 1.10 Different types of X-ray imaging systems. (A) The image intensifier consists of an input phosphor coupled to a photocathode which converts X-rays into optical photons and subsequently electrons. The latter are accelerated in an electric field and hit an output phosphor screen. This produces an amplification of several thousand. The resulting optical image is captured using an optical assembly and a CCD. (B) The indirect active matrix flat panel imager (AMFPI) consists of a scintillator which converts X-rays into optical photons, a photoconductor which converts them in turn to electrons and a readout layer which stores and processes the resulting charge image. (C) In the direct conversion flat-panel detector, X-rays are directly converted into charge inside a photoconductor. In each diagram, the electric field lines are shown as two lines next to an arrow which shows the direction in which the charge travels.

which small numbers of X-rays—ideally a single X-ray—can be detected) due to the large internal gain of the XRII, however, the XRII also presents important limitations. These are associated with the curvature of the input phosphor and the presence of multiple conversion stages, leading to geometrical distortions (particularly in the periphery of the images), spatial nonuniformities, and a degradation of the imaging resolution. Furthermore, these systems are bulk and heavy (they can weigh several hundred pounds), compromising patient accessibility and image acquisition modes such as rotational angiography or cone-beam CT. Another important limitation of the XRII is its high sensitivity to magnetic fields (including the earth's magnetic field) which produces a distortion in the image that is shaped like an "S" and referred to as an S-distortion.

Due to these substantial limitations and the availability of new solid-state technologies, there has been a progressive trend in the past few years to replace XRII systems with flat-panel detectors.

1.5.5 A New Revolution in Optical Imaging

Although improvement in recording media considerably enhanced the quality and use of X-rays for medical purposes, with the notable exception of fluoroscopy and nuclear imaging developed decades later, it was not until the early 1970s that a new revolution began in medical imaging.

The era of modern medical imaging did not begin until the 1970s. This modern era was heralded again by an X-ray imaging device called the "CAT" scanner or Computerized Axial Tomograph device, which has long since been called simply Computed Tomography, or CT. This device contributed three major significant features to medical imaging that continues to be its foundation today: (i) the image was digital, produced by a computer, and could be readily modified, analyzed, and displayed by computers; (ii) the method provided sensitivity to tissue density differences unattainable theretofore; and (iii) the methodology provided cross-sectional views of the human body, eventually multiple cross-sectional views adjacent one to another, providing a 3D view of internal anatomic structures. Since the body is 3D and the organs have different shapes, sizes, and positions relative one to another in 3D space, imaging adjacent thin body cross sections was an important advance in medicine's ability to more accurately see and understand the true nature of objects inside the body.

Soon after the advent of this scanner in the 1970s, 3D imaging became available and other modalities began to be rapidly developed. Particular note is given to MRI in the early 1980s because of its near revolutionary impact on soft tissue imaging. Nuclear systems, notably positron emission tomography (PET), introduced functional imaging, followed by high-speed CT (helical) and rapid MRI (fMRI). By 1990s, with significant performance gains in imaging methodologies, interactive multidimensional, multimodality imaging for improved diagnosis, treatment, guidance, and therapy monitoring became routine in large medical centers. The turn to the twenty-first century was characterized by the advent of many image-guided interventions, often associated with minimally invasive surgery, for planning, rehearsal, and execution of a wide variety of clinical and surgical procedures.

The continuing goals for development and acceptance of important visualization display technology are: (i) improvement in speed, quality, and dimensionality of the display and (ii) improved access to the data represented in the display through interactive, intuitive manipulation, and measurement of the data represented by the display. Included in these objectives is determination of the quantitative information about the properties of anatomic tissues and their functions that relate to and are affected by disease. With these advances in hand, the delivery of several important clinical applications will soon be possible. Major events in diagnostic imaging are given in Table 1.3.

These significant advances have set the stage for an exciting future that will include highly sensitive and specific imaging, real-time, and multidimensional imaging, whereby almost any number of multiple orthogonal image variables can be fused and synchronized together to bring all collected information synergistically to bear on diagnosis and treatment of disease. The near future will demonstrate highly integrated capabilities for structural and functional information synchronously across space and time, and this will drive the practice of medicine of the future toward truly synchronous, minimally invasive, highly specific, highly sensitive, and highly effective diagnosis and treatment of disease.

Table 1.3 Major Events in Diagnostic Imaging over the Last Five Decades

1950s	Utilization of an image-intensifier system for fluoroscopy
	Development of a gamma camera for radionuclide imaging
1960s	Development of a 90 s automated film processor
	Basic research on image quality, MTF, Weiner spectra, and quantum mottle
1970s	Development of rare-earth screen–film system, DSA, ultrasound imaging with electronic scan
	Initial research on ROC analysis, MRI, PET, SPECT, PACS, and electronic imaging
1980s	Development of CR, MRI, color Doppler ultrasound imaging
	Initial research on computer-aided diagnosis (CAD)
	Commercialization and clinical use of CAD system, flat-panel detector (FPD) systems, multidetector computed tomography (MDCT), magnetic resonance angiography (MRA), harmonic/contrast imaging
1990s	Development and clinical use of real-time 3D ultrasound imaging, cone-beam CT, parallel MRI PET/CT, full-field digital mammography (FFDM), molecular imaging and PACS

Avalanche multiplication of charge in a-Se is the only that can provide sufficient gain to satisfy requirement (1) and the highly adjustable avalanche gain should also satisfy requirement (2). Furthermore, the high intrinsic imaging resolution of a-Se should also answer requirement (3) [19,20].

1.6 Risk Factor

Exposure to X-rays leaves very small residual amounts of radiation in the subject. These build up cumulatively over a lifetime, causing cancer at high-enough levels. Because it does take a long time for this damage to happen, X-rays were not immediately suspected as the cause. The first recorded death from X-ray radiation damage was Clarence Dally, one of Thomas Edison's assistants. Through the first half of the twentieth century, intensive research was done into the effects of this radiation, and protective measures (such as lead shielding) developed to reduce exposure to it.

Undoubtedly, the early detection of health problems is very important. Over the last decades, various imaging techniques such as X-rays, CT, and magnetic resonance scans have been developed and applied to diagnostic, as well as therapeutic, medical care.

However, in recent years, physicians have become concerned with the overuse of certain diagnostic techniques, in particular those that expose patients to radiation. Although infrequent use of X-ray or CT scans will not have adverse effects on a patient, multiple exposures to radiation over a short period of time can cause serious damage to cells, resulting in an increased risk of cancer and other diseases.

The harmful effects of ionizing radiation used in fluoroscopy, which have long been recognized, require the patient dose to be as low as reasonably achievable during an intervention. This is often referred to as the ALARA principle. The use of harmful X-ray radiation is justifiable by considering that the benefit from the clinical outcome of the intervention will outweigh the adverse biological effects of the radiation.

These biological effects include indirect and direct effects.

Indirect effects of ionizing radiation arise when electrons set in motion by X-ray photons excite and ionize water molecules, creating free radicals which then cause damage to critical biological targets such as DNA.

In **direct effects**, electrons directly ionize DNA molecules. As a result, in certain cases such as pediatric interventions, a particularly strict adherence to ALARA is required, since the accrued stochastic effects due to radiation exposure are more likely to disrupt tissue growth and development as well as lead to an increased chance of cancer over the child's lifetime. There are adverse effects associated with other imaging modalities as well. The electromagnetic radiofrequency pulses used in MRI are known to cause heating. This can be particularly problematic near metallic devices or implants such as pacemakers or hearing aids. Ultrasound contrast agents, when exposed to ultrasound waves, can also cause potential bioeffects (i.e., rupture of cell membranes) at the level of the microcirculation, although the clinical relevance of such bioeffects remains unclear.

1.7 Medical Exposure Dosimetry

The magnitudes of equivalent dose and effective dose are radiation protection terms for stochastic effects used by the International Commission on Radiological Protection (ICRP) in prospective dose assessments for planning and optimization in radiological protection, and for demonstration of regulatory compliance with dose limits of occupationally exposed workers and members of the public. Their unit is the sievert (Sv). "Effective dose is not recommended for epidemiological evaluations, nor should it be used for detailed specific retrospective investigations of individual exposure and risk [18]." In medical exposure, effective dose can be used as a quality control tool to compare doses from different radiation modalities, as shown in table 6 from AAPM Report 96 [18] but not to assess individual patient doses. The International Commission on Radiation Units and Measurements (ICRU) states that "to assess the risk from stochastic and deterministic effects from medical x-ray imaging, it is necessary to know the organ or tissue doses, the dose distribution and the age and gender of the patients." The quantities and units to be used in medical X-ray imaging as well as methods for patient dose calculation and measurements are given in ICRU Report. The unit is the gray (Gy).

To determine patient doses in radiography/fluoroscopy, one can measure the kerma in air for all radiation beams and multiply it by the technique factors used, collected either manually or from the DICOM header. Another possibility is to attach to each X-ray machine a properly calibrated air-kerma-area-product meter (previously called DAP meter). CT doses may be calculated from "computed tomography air-kerma (dose) index" (CTDI) and "air-kerma (dose)—length product" (DLP) measurements [15–18]. Doses can also be measured directly by placing thermoluminiscent dosimeters or diodes on the patients during the procedures (Table 1.4). The dosimetric role of the medical physicist is crucial.

According to the 2008 United Nations Scientific Committee on the Effects of Atomic Radiation (UNSCEAR) report [18], approximately 3.6 billion diagnostic (3.1 medical and 0.5 dental) X-ray examinations are undertaken annually in the world. There is a significant increase in annual examination frequency for this time period (1997–2007) in comparison with previous UNSCEAR surveys. This is mainly due to the aging of the population, as most medical exposures are performed on older individuals. Regarding dose, as the table shows, the global annual dose from diagnostic medical exposures is 0.61 mSv, which represents 20% of the world total, estimated as 3.1 mSv (natural background contributes 79%). This percentage contribution is very different in industrialized countries such as the United States where the rise in medical uses in the period 1980–2006 has resulted in an increase in the total annual effective dose from 3.0 to 5.4 mSv, making medical exposure equal to or larger than exposure due to natural background. The figure will increase further if mass screening techniques with CT, currently under clinical trails, such as lung cancer screening for smokers and ex-smokers, calcium scoring for atherosclerosis, coronary angiography, and virtual colonoscopy, show benefits.

Reasonably, the first step in radiological protection of diagnostic medical exposures is to assess whether the X-ray procedure is really needed, i.e., will the study affect patient management.

In diagnostic medical exposures, patient dose has to be kept to the minimum necessary to achieve the required diagnostic objective, taking into account norms of acceptable image quality established by appropriate professional bodies and relevant reference levels for medical exposure. This means that exposures resulting in doses above clinically acceptable minimum doses must be avoided. However, it is important to understand that optimization of protection does not mean dose reduction, and that diagnostic information, not image quality, should be the deciding factor. Optimized images have to be established based on the characteristics of the

Table 1.4 Typical Effective doses for Several Common Imaging Exams

Radiography/Fluoroscopy (mSv)		**CT (mSv)**	
Hand radiograph	<0.1	Head CT	1–2
Dental bitewing	<0.1	Chest CT	5–7
Chest radiograph	0.1–0.2	Abdomen CT	5–7
Mammogram	0.3–0.6	Pelvus CT	3–4
Lumbar spine radiograph	0.5–1.5	Abdomen and pelvus CT	8–14
Barium enema exam	3–6	Coronary artery calcium CT	1–3
Coronary angiogram (diagnostic)	5–10	Coronary CT angiography CT	5–15

image receptor, the patient habitus and the purpose of the radiological examination. Particular attention should be paid to children and pregnant females.

Since the relationship between exposure and noise is an inverse relationship, noise is the major factor that limits how much X-ray exposure and patient dose can be reduced. This applies to all forms of X-ray imaging, including fluoroscopy and CT. In S/F radiography, the film serves as the acquisition, display, and archival medium and thus, the process must be optimized during the acquisition phase of the image production. In DR, on the other hand, the three processes are separated and image optimization can be accomplished by pre- and postprocessing software. Noise reduction methods other than increasing the photon fluence have been published [15–18].

1.8 Therapeutic Medical Physics

The employment of physicists in radiation therapy dates back to soon after X-rays were first used for treatment of diseases in the late 1890s and early 1900s. Physicists were needed because the early X-ray machines required constant nurturing to keep them running reliably and with some consistency in dose delivery. Many of the early developments in dose specification and measurement were made by physicists, culminating in the establishment of the first unit of "dose," the roentgen, in 1928. During this same period, a number of hospitals began to employ physicists to deal with the handling and dosimetry associated with radium and radon brachytherapy. Two pioneers in the United States whose names come to mind readily in this context were Giaoacchino Failla, DSc, and Edith Quimby, DSc, at the Memorial Hospital in New York City [17].

Medical physics has changed dramatically since 1895. There was a period of slow evolutionary change during the first 70 years after Roentgen's discovery of X- rays. With the advent of the computer, however, both diagnostic and therapeutic radiology have undergone rapid growth and changes [17]. Technologic advances such as CT and MRI in diagnostic imaging and 3D treatment planning systems, stereotactic radiosurgery, and intensity-modulated radiation therapy in radiation oncology have resulted in substantial changes in medical physics. These advances have improved diagnostic imaging and radiation therapy while expanding the need for better educated and experienced medical physics staff.

References

[1] O.W. Linton, Am. J. Roentgenol. 165 (1995) 471.
[2] A. Assmus, Beamline 10 (1995).
[3] R.F. Mould, Phys. Med. Biol. 40 (1995) 1741.
[4] W.C. Roentgen, Zitzungsberichte der Wurzburger Physik-medico Gegellshraft (1895).
[5] Scientific American 74 (1896) 51.

[6] Science, 3 (1896) 161.
[7] W.C. Roentgen, Nature 53 (1896) 274.
[8] W. Friedrich, P. Knipping, M. Laue, Sitzungber Bayer Akad Wiss (1912) 303.
[9] M. von Laue, Sitzungber Bayer Akad Wiss 363 (1912).
[10] M. von Laue, Concerning the detection of X-ray interferences, Nobel Lecture, December 12, 1915. Available from: www.nobelprize.org.
[11] W.W. Schmahl, W. Steurer, Acta Cryst. A68 (2012) 1.
[12] W.L. Bragg, Proc. Cambridge Philos. Soc. 17 (1913) 43.
[13] J. Gribbin, The Scientists. A History of Science Told Through the Lives of its Greatest Inventors, Random House, New York, NY, 2004.
[14] M. Bradaczek, M. Popescu, J. Optoelectron. Adv. Mater. 9 (7) (2007) 1945.
[15] Radiography in Modern Industry, Eastmen Kodak (1980).
[16] GE Inspection Technologies, General Electric Company (2007).
[17] J.E. Gray, C.G. Orton, Radiology 217 (2000) 621.
[18] C. Borras, Radiation protection in diagnostic radiology, Rev. Panam Salud Publica/Pan Am. J. Public Health 20 (2006) 1.
[19] S.O. Kasap, J.B. Frey, G. Belev, O. Tousignant, H. Mani, J. Greenspam, et al., Sensors 11 (2011) 5112.
[20] J.A. Rowlands, Phys. Med. Biol. 47 (2002) R123.

2 Computed Tomography

2.1 Definition of CT; Important Terms

CT is a diagnostic procedure that uses special X-ray equipment to create cross-sectional pictures of human body.

Absorption: Some of the incident X-ray energy is absorbed in patient tissues and hence does not contribute to the transmitted beam.

Anode: Tungsten bombarded by a beam of electrons to produce X-rays. In all but one-fifth generation system, the anode rotates to distribute the resulting heat around the perimeter. The anode heat-storage capacity and maximum cooling rate often limit the maximum scanning rates of CT systems.

Attenuation: The total decrease in the intensity of the primary X-ray beam as it passes through the patient, resulting from both scatter and absorption processes. It is characterized by the linear attenuation coefficient.

Control console: The control console is used by the CT operator to control the scanning operations, image reconstruction, and image display.

Cormack, Dr. Allan MacLeod: A physicist who developed mathematical techniques required in the reconstruction of tomographic images. Dr. Cormack shared the Nobel Prize in Medicine and Physiology with Sir Dr. G. N. Hounsfield in 1979.

Data-acquisition system (DAS): Interfaces the X-ray detectors to the system computer and may consist of a preamplifier, integrator, multiplexer, logarithmic amplifier, and analog-to-digital converter.

Detector array: It is an array of individual detector elements. The number of detector elements varies between a few hundred and 4800, depending on the acquisition geometry and manufacturer. Each detector element functions independently of the others.

Fan beam: The X-ray beam is generated at the focal spot and so diverges as it passes through the patient to the detector array. The thickness of the beam is generally selectable between 1.0 and 10 mm and defines the slice thickness.

Focal spot: Briefly, it can be considered as the region of the anode where X-rays are generated.

Focused septa: Thin metal plates between detector elements which are aligned with the focal spot so that the primary beam passes to the detector elements without losses, while scattered X-rays which normally travel in an altered direction are blocked.

Gantry: It is the largest component of the CT installation. This component include: the X-ray tube, collimators, detector array, DAS, other control electronics, the mechanical components required for the scanning motions.

Helical scanning: The scanning motions in which the X-ray tube rotates continuously around the patient while the patient is continuously translated through the fan beam. The focal spot therefore traces a helix around the patient. Projection data are obtained which

Medical Imaging Technology. DOI: http://dx.doi.org/10.1016/B978-0-12-417021-6.00002-2

allow the reconstruction of multiple contiguous images. This operation is sometimes called spiral, volume, or three-dimensional CT scanning.
Sir Hounsfield, Dr. Godfrey Newbold: A physicist who developed the first practical CT instrument in 1971. Dr. Hounsfield received the McRobert Award in 1972 and shared the Nobel Prize in Medicine and Physiology with Dr. A. M. Cormack in 1979 for this invention.
Image plane: It is the plane through the patient that is imaged. In practice, this plane (also called a slice) has a selectable thickness between 1.0 and 10 mm centered on the image plane.
Pencil beam: A narrow, well-collimated beam of X-rays.
Projection data: The set of transmission measurements used to reconstruct the image.
Reconstruct: The tomographic image obtained from the projection data by the help of the mathematical operation.
Scan time: The time required to acquire the projection data for one image, typically 1.0 s.
Scattered radiation: Radiation caused by that is removed from beam by a scattering process. This radiation is not absorbed and continues along a path in an altered direction.
Slice: See Image plane.
Spatial resolution: Spatial resolution is a measure of how well a detail is delineated in an image. The spatial resolution in the scan plane is referred to as the axial spatial resolution. Resolution in the coordinate perpendicular to the scan plane, i.e., in the z-direction, is referred to as the longitudinal spatial resolution.
Spiral scanning: See Helical scanning.
Three-dimensional imaging: See Helical scanning.
Tomography: A technique used for obtaining a cross-sectional slice images.
Volume CT: See Helical scanning.
X-ray detector: A device used for absorbing X-ray radiation and converting some or all of the absorbed energy into a small electrical signal.
X-ray linear attenuation coefficient μ: Expresses the relative rate of attenuation of a radiation beam as it passes through a material. The value of μ depends on the density and atomic number of the material and on the X-ray energy. The unit of μ is cm^{-1}.
X-ray source: This device used for X-ray beam generation. All CT scanners are rotating-anode bremsstrahlung X-ray tubes except one-fifth generation system, which uses a unique scanned electron beam and a strip anode.
X-ray transmission: The fraction of the X-ray beam intensity that is transmitted through the patient without being scattered or absorbed. It is equal to I_t/I_0 and can be can be determined by measuring the beam intensity both with (I_t) and with ought (I_0) the patient present. As a rule of thumb, n^2 independent transmission measurements are required to reconstruct an image with an $n \times n$ sized pixel matrix.

2.2 A Brief History of CT

In a case of conventional planar projected images of the patient, some important details remained to be hidden by overlaying tissues. In contrast, by using slice-imaging techniques (tomography), selective demonstration of morphologic properties becomes available, layer by layer, and additional, vitally important information can be extracted [1–5].

CT imaging is usually known as "CAT scanning" (computed axial tomography). The term "tomography" is from the Greek word "*tomos*" meaning "slice" or "section" and "*graphia*" meaning "describing." In 1972, British engineer Godfrey Hounsfield and American physicist Allan M. Cormack had invented CT by combining computer technology with X-ray technology. Notably that Hounsfield's discovery was all the more astonishing because he had previously worked for the British record company External Machine Interface (EMI) which also manufactured electronic components.

Computerized tomography is nearly an ideal form of tomography yielding sequence images of thin consecutive slices of the patient. Moreover, CT provides the opportunity to localize resultant image in three dimensions. Unlike conventional, classical tomography, computerized tomography does not suffer from interference from structures in the patient outside the slice being imaged. This is achieved by irradiating only thin slices of the patient with a fan-shaped beam.

Transaxial images (tomograms) of the patient's anatomy can give more selective information than conventional planar projection radiographs. Compared to planar radiography, CT images have superior contrast resolution, i.e., they are capable of distinguishing very small (even vanishingly small) differences in tissue-attenuation (contrasts). At the same time, they have inferior spatial resolution. An attenuation difference of 0.4% can be visualized but the smallest details in the image that can be resolved must be separated by 0.5 mm at least. In conventional planar radiography, the lowest detectable contrast is larger but details of smaller size can be separated.

CT images are produced using X-ray technology and powerful computers.

X-ray CT involves several disciplines: *physics*, *mathematics*, *computer science*, *technology*, and *medicine*, as well as the spectrum of personalities involved in those disciplines. Each of these disciplines evolved in its own right, but only occasionally was the connection between the developments in the different disciplines realized or acknowledged. The mathematical basis is crucial, but the effective mathematical methods turn out to be only practically useable with a programmable digital computer. Such machines, suitable for the large-scale computations involved in CT, did not become generally available until the 1960s, when the computer chip was developed [9]. Thus, the Dutchman Bockwinkel developed a mathematical approach to solve a crystallographic problem that could well have been used as the basis of CT in the late 1900s. The Austrian Radon (1917) had little direct impact on the initial development of X-ray CT imaging, even though Radon's Transform turned out to be a more direct match to the needs of CT. Only after aspects of that Transform were developed independently some 60 years later was it realized that Radon had already published this approach.

The first computer in the world to be officially recognized as such is the ENIAC machine (Electronic Numerical Integrator and Calculator) from 1946. The existence of the computer, called Colossus I, was kept secret until 1976. It is worth remembering that the first computers were far from perfect. They contained about 18 thousand very unreliable valves; this meant that the time that the computers were out of commission considerably exceeded the time that they worked. Over the years, engineering advances and progress toward the miniaturization of components in computers led to the development of microcomputers.

The two people generally credited with inventing CT were awarded the Nobel Prize for Physiology and Medicine in 1979: Allan MacLeod Cormack (1924–1998) and Godfrey Newbold Hounsfield (1919–2004). Although the Norwegian Abel conceived the idea of tomography significantly earlier (in 1826), and then the Austrian Radon developed it further, it was only the solution proposed by Cormack and Hounsfield that fully deserves the name computed tomography.

Cormack published the results of his research in an article entitled "Representation of a Function by its Line Integrals." As a theoretical physicist, Cormack was not concerned about the practical application of his research. It was the work of the Englishman Godfrey Newbold Hounsfield, employed at the Central Research Laboratories of EMI Ltd., which led to the construction of the first CT scanner.

It is also interesting to note that the two people, who are recognized by historians of science as the fathers of CT (Figure 2.1), only met each other for the first time in 1979 at the presentation ceremony, where they jointly received the Nobel Prize in Physiology and Medicine.

The first clinical CT scanners were installed between 1974 and 1976. The original CT systems were dedicated to head imaging only. The "whole body" systems with larger patient openings became available in 1976. CT became widely available by about 1980. There are now more than 6000 CT scanners installed in the United States and about 30,000 installed worldwide.

The first CT scanner developed by Hounsfield in his laboratory at EMI took several hours to acquire the raw data for a single scan or "slice" and took days to reconstruct a single image from this raw data. The latest multislice CT systems can collect up to 4 slices of data in about 350 ms and reconstruct a 512×512 matrix image from millions of data points in less than a second. An entire chest (forty 8 mm slices) can be scanned in 5–10 s using the most advanced multislice CT system.

During its 35-year history, CT has made great improvements in speed, patient comfort, and resolution. As CT scan times have gotten faster, more anatomy can be scanned in less time. Faster scanning helps to eliminate artifacts from patient motion such as breathing or peristalsis. CT exams are now quicker and more patient-friendly than ever before. Tremendous research and development has been made to provide excellent image quality for diagnostic confidence at the lowest possible X-ray dose.

Figure 2.1 Allan MacLeod Cormack (left) and Godfrey Newbold Hounsfield (right).

2.3 Principles of Operation and the Procedure

To derive CT image, two steps are necessary [1–5]:

> Firstly, physical measurements of the attenuation of X-rays traversing the patient in different directions.
> Secondly, mathematical calculations of linear attenuation coefficients, μ, all over the slice are necessary.

The procedure is as follows. The patient remains stationary on the examination table. The X-ray tube rotates in a circular orbit around the patient in a plane perpendicular to the length-axis of the patient. A fan-shaped beam of variable thickness (1–10 mm), wide enough to pass on both sides of the patient is used. The X-ray tube is similar to but more powerful than those used in planar radiography. The image receptor is an array of several hundred small separate receptors. Readings from the receptors are fed in to a computer. The later, after numerous calculations, produces a tomogram of the patient, in other words gives a map of linear attenuation coefficients μ [6–8].

Early CT scanners acquired images a single slice at a time (the procedure is known as sequential scanning). During the 1980s, significant advancements in technology heralded the development of slip ring technology. In these CT machines, the X-ray tube rotates continuously in one direction around the patient. This has contributed to the development of helical or spiral CT.

In spiral CT, the X-ray tube rotates continuously in one direction while the table on which the patient is lying is mechanically moved through the X-ray beam. The transmitted radiation thus takes on the form of a helix or spiral. Instead of acquiring data one slice at a time, information can be acquired as a continuous volume of contiguous slices. This allows larger anatomical regions of the body to be imaged during a single breath hold, thereby reducing the possibility of artifacts caused by patient movement. Faster scanning also increases patient throughput and increases the probability of a diagnostically useful scan in patients who are unable to fully cooperate with the investigation. Among other advantages of spiral CT, it should be noted for its speed and continuity. There is no doubt that advantages mentioned earlier are important when CT is applied for obtaining imaging of the lungs. The speed allows the complete data set to be acquired within one breath hold. This avoids the problem of parts of the lung being missed because the patient breathes differently in consecutive acquisitions.

The speed of acquisition also makes it possible to make better use of the contrast dynamics of intravenously administered contrast agents, e.g., in CT angiography (CTA). This allows blood vessels to be imaged while they contain the maximum amount of contrast agent. It is also possible to obtain images of a complete organ in one or several specific phases of contrast.

These are the dominant type of scanners on the market because they have been manufactured longer and offer lower cost of production and purchase.

The main limitation of spiral scanners is that they are bulk. Inertia of the equipment also should be mentioned—X-ray tube assembly and detector array on the opposite side of the circle limits the speed at which the equipment can spin. Figures 2.2–2.5 illustrate the evolution of CT machines.

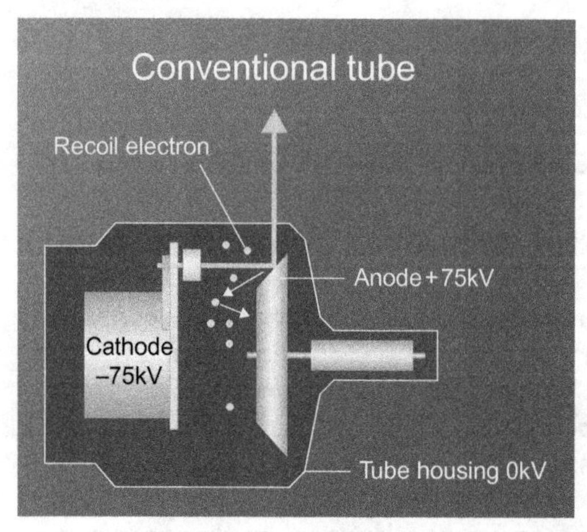

Figure 2.2 Conventional X-ray tube (collected by Afnan Malaih/Lamis Jada'a) www.kau.edu.sa.

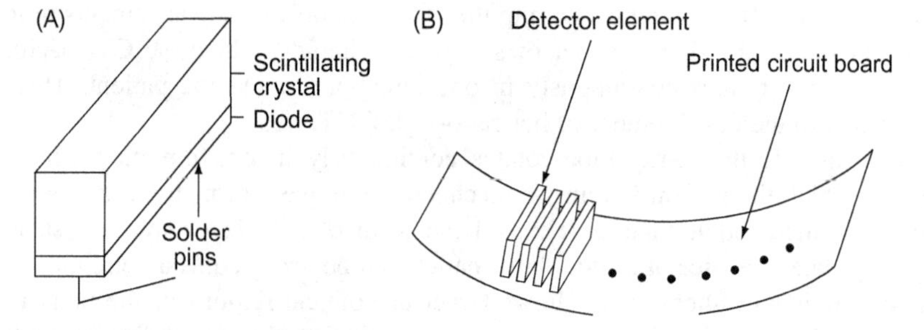

Figure 2.3 (A) A solid-state detector consists of a scintillating crystal and photodiode combination. (B) Many such detectors are placed side by side to form a detector array that may contain up to 4800 detectors.

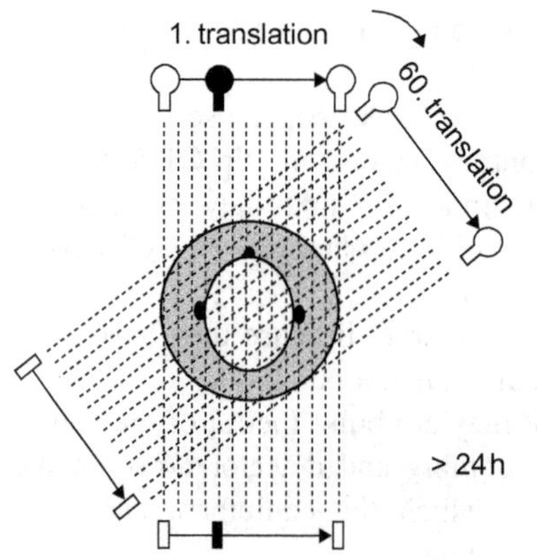

Figure 2.4 First generation parallel-beam geometry, (translation/rotation).

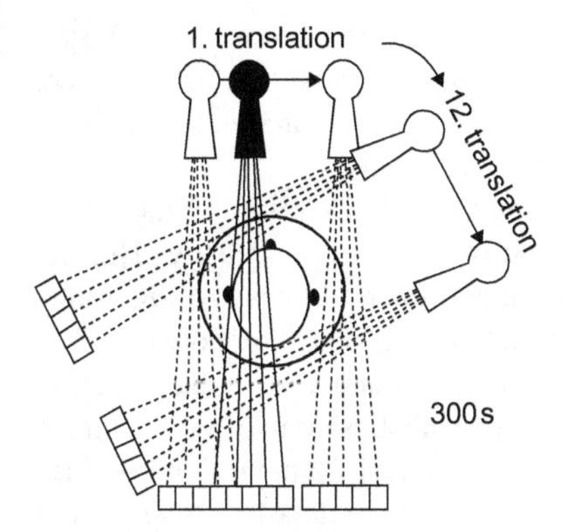

Figure 2.5 Second generation fan beam (multiple detectors).

2.4 CT Equipment

2.4.1 CT Gantry

The scan (or imaging) system is the major component of any CT system. This primarily includes: the gantry; the patient table (also known as couch). The gantry is a moveable frame that contains the X-ray tube including collimators, filters, detectors, DAS, and rotational components. Rotational components, in turn, include slip ring systems and all associated electronics such as gantry angulation motors and positioning laser lights. In earlier CT systems, a power to the X-ray tube and the rotational components is supplied by small generator via cables for operation. In some cases, generator was mounted on the rotational component of the CT system and rotated together with the X-ray tube. In other cases, generators remain mounted inside the gantry wall. Newer scanner designs utilize a generator that is located outside the gantry. Slip ring technology eliminated the need for cables and allows continuous rotation of the gantry components. The inclusion of slip ring technology into a CT system allows for continuous scanning without interference of cables. A CT gantry can be angled up to 30° toward a forward or backward position. Gantry angulation is determined by the manufacturer and varies among CT systems. Gantry angulation allows the operator to align pertinent anatomy with the scanning plane. The opening through which a patient passes is referred usually to as the gantry aperture. Gantry aperture diameters generally range from 50 to 85 cm. Generally, larger gantry aperture diameters, 70–85 cm, are necessary for CT departments which do a large volume of biopsy procedures. The larger gantry aperture allows practitioners more easily to manipulate with CT equipment at the biopsy procedure and reduces the risk of injury.

2.4.2 X-Ray Tube

The important variables that determine the efficiency of each X-ray source for a particular task are:

a. The size of the focal spot
b. The spectrum of X-ray energies generated
c. The X-ray intensity.

The focal-spot size partially defines the potential spatial resolution of a CT system by determining the number of possible source-detector paths that can intersect a given point in the object being scanned. The more such source-detector paths there are, the more blurring of features there will be observed.

The energy spectrum defines the penetrative ability of the X-rays and their relative attenuation as they pass through materials/body of different density. Higher-energy X-rays penetrate more effectively than lower-energy ones. At the same time, they are less sensitive to changes in material density and composition. The X-ray intensity directly affects the signal-to-noise ratio and thus image clarity. Higher intensities improve the underlying counting statistics, but often require a larger focal spot. Many conventional X-ray tubes have a dual filament that provides two focal-spot sizes, with the smaller spot size allowing more detailed imagery at a cost in intensity. Medical CT systems tend to have X-ray spot sizes that range from 0.5 to 2 mm. The high-resolution system at the Universal Time (UT) CT facility utilizes a dual-spot 420 kV X-ray source (Pantak HF420), with spot sizes of 0.8 and 1.8 mm. The small spot has a maximum load of 800 W (i.e., 2 mA at 400 kV).

The large spot has a maximum load of 2000 W. The 225 kV tube used for ultra-high resolution work (Fine focus FXE-225.20) has an adjustable focal spot with a minimum size of $<6\ \mu m$ at 8 W total load, but at higher loads the spot size is automatically increased to prevent thermal damage to the target. In most cases, a slightly "defocused" beam (larger spot size) can be used to improve counting statistics with little cost in resolution. Both sources have tungsten targets.

X-ray tubes may all be classified in two general groups:

1. Gas tubes
2. High-vacuum tubes

As for the first ones, they are freed from a cold cathode by positive ion bombardment. For the existence of the positive ions a certain gas pressure is required without which the tube will allow no current to pass. Metals, such as platinum and tungsten, are placed in the path of the electron beam to serve as the target. Concave metal cathodes are used to focus the electrons on a small area of the metal target and increase the sharpness of the resulting shadows on the fluorescent screen or the photographic film. Many designs of gas tubes have been built for medical applications.

In the case of high-vacuum tubes, one faced with the operational difficulties and erratic behavior of gas X-ray tubes are inherently associated with the gas itself and the positive ion bombardment. The latter takes place during operation. The high-vacuum X-ray tube eliminates these difficulties—other means of emitting electrons from the cathode. Originally, high-vacuum X-ray tube had a hot tungsten-filament cathode and a solid tungsten target. This tube permitted stable and reproducible operation with relatively high voltages and large masses of metals. A modern, commercially available, hot-cathode high-vacuum X-ray tube is built with a liquid-cooled, copper-backed tungsten target.

To make the cathode and electron beam target, tungsten is used. Among other materials which can be found in the tube are: Pyrex, glass, copper, and tungsten alloys. Throughout many parts of the CAT scanner system, lead can be found. This material reduces the amount of excess radiation.

Constructively, the X-ray tube is made like other types of electrical diodes. The individual components, including the cathode and anode, are placed inside the tube envelope and vacuum sealed. The tube is then situated into the protective housing, which can then be attached to the rotating portion of the scanner frame.

To create the large amount of voltage needed to produce X-rays, an autotransformer is used. This power supply device is made by winding wire around a core. Electric tap connections are made at various points along the coil and connected to the main power source. With this device, output voltage can be increased to approximately twice the input voltage.

As in a case of any vacuum tube, there is a cathode. The latter emit electrons into the vacuum. At the same time, the function of an anode is to collect the electrons, thus establishing a flow of electrical current through the tube. A high-voltage power source is connected across cathode and anode, e.g., 30–150 kV. Thus, the voltage can be quickly switched on, then off, for precise amounts of time, e.g., 0.001–1.0 s. One can control the current flow, the 1.0–1000 mA range, once it started.

Until the late 1980s, X-ray generators were merely high-voltage, AC to DC variable power supplies. In the late 1980s, a different method of control was emerging, called high-speed switching. This followed the electronics technology of switching power supplies (switch mode power supply) and allowed for more accurate control of the X-ray unit, higher-quality results, and reduced X-ray exposures.

Electrons focused on the tungsten (or sometimes molybdenum) anode from the cathode, collide with and accelerate other electrons, ions, and nuclei within the anode material. As a result, the energy (approximately 1% of generated energy) is emitted/radiated, perpendicular to the path of the electron current, as X-ray photons. In addition, about 1% of the energy generated is emitted/radiated, perpendicular to the path of the electron current, as X-ray photons. Over time, tungsten will be deposited from the anode onto the interior glass surface of the tube. This will slowly degrade the quality of the X-ray beam. The tungsten deposit seemed to be sufficient enough to act as a conductive bridge. That is, at high enough settings, arcing will occur. The arc will jump from the cathode to the tungsten deposit, and then to the anode. This arcing will cause an effect called "crazing" on the interior glass of the X-ray window. As time goes on, the tube will become unstable even at lower potentials and must be replaced. At this point, the tube assembly (also called the "tube head") is removed from the X-ray system and replaced with a new tube assembly. The old tube assembly is shipped to a company that reloads it with a new X-ray tube.

The X-ray photon-generating effect is generally called the bremsstrahlung effect, a contraction of the German *brems* for braking, and *strahlung* for radiation.

The range of photonic energies emitted by the system can be adjusted by changing the applied voltage and installing aluminum filters of varying thicknesses. Aluminum filters are installed in the path of the X-ray beam to remove "soft" (non-penetrating) radiation.

The numbers of emitted X-ray photons, or dose, are adjusted by controlling the current flow and exposure time. In other words, the high-voltage controls X-ray penetration, and thus the contrast of the image. The tube current and exposure time affect the dose and therefore the darkness of the image.

It should be noted here that X-ray tubes are subjected to far higher thermal loads in CT than in any other diagnostic X-ray application. In early CT scanners, such as in first- and second-generation, stationary anode X-ray tubes were used, since the long scan times meant that the instantaneous power level was low. Long scan times also allowed heat dissipation. Shorter scan times in later versions of CT scanners required high-power-ray tubes. The use of oil-cooled rotating anodes needed for efficient thermal dissipation. Heat-storage capacities varied from 1 to 3 million heat units in early third-generation CT scanners. The introduction of helical CT with continuous scanner rotation placed new demands on X-ray tubes. Several technical advances in component design have been made to achieve these power levels and deal with the problems of target temperature, heat storage, and heat dissipation. For example, the tube envelope, cathode assembly, and anode assemblies including anode rotation and target design have been redesigned. As scan times have decreased, anode heat capacities have increased by as much as a factor of five, preventing the need for cooling delays during most clinical procedures, and tubes with capacities of 5–8 million heat units are available. In addition, improvement in the heat dissipation rate (kilo-heat units per minute) has increased the heat-storage capacity of modern X-ray tubes. The large heat capacities are achieved with thick graphite backing of target disks (anode diameters of 200 mm or more) improved high temperature rotor bearings and metal housings with ceramic insulators among other factors.

The "lifetime" of tubes used to date ranges from 10,000 to 40,000 h, while for compared with the 1000 h. Because many of the engineering changes increased the mass of the tube, much of the design effort was also dedicated to reduce the mass to better withstand increasing gantry rotational rates required by faster scan times.

All CT scanners with the exception of latest—third, fourth (Figures 2.6 and 2.7), and fifth generations—use bremsstrahlung X-ray tubes as the source of radiation. These tubes are typical of those used in diagnostic imaging and produce X-rays by accelerating a beam of electrons onto a target anode. The anode area, from which X-rays are emitted, projected along the direction of the beam, is called the focal spot. Most systems have two possible focal spot sizes,

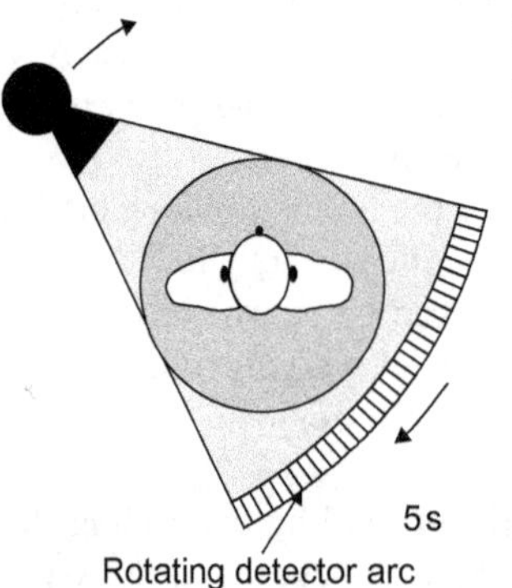

Figure 2.6 Third generation fan beam (continuous rotation).

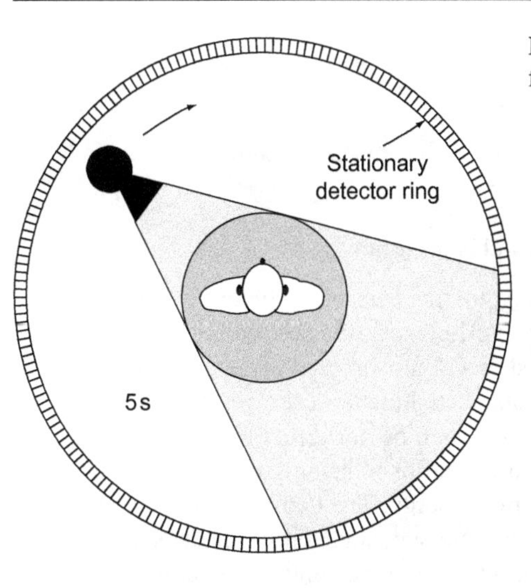

Figure 2.7 Fourth generation fan beam, fixed detectors.

approximately $0.5 \times 1.5\ mm^2$ and $1.0 \times 2.5\ mm^2$. A collimator assembly is used to control the width of the fan beam between 1.0 and 10 mm, which in turn controls the width of the imaged slice.

The power requirements of these tubes are typically 120 kV at 200–500 mA, producing X-rays with an energy spectrum ranging between approximately 30 and 120 keV. All modern systems use high-frequency generators, typically operating between 5 and 50 kHz. Some spiral systems use a stationary generator in the gantry, requiring high-voltage (120 kV) slip rings, while others use a rotating generator with lower-voltage (480 V) slip rings.

Production of X-rays in bremsstrahlung tubes is an inefficient process, and hence most of the power delivered to the tubes results in heating of the anode. A heat exchanger on the rotating gantry is used to cool the tube. Spiral scanning, in particular, places heavy demands on the heat-storage capacity and cooling rate of the X-ray tube.

The intensity of the X-ray beam is attenuated by absorption and scattering processes as it passes through the patient. The degree of attenuation depends on the energy spectrum of the X-rays as well as on the average atomic number and mass density of the patient tissues. The transmitted intensity is given by:

$$\mathrm{I} = \mathrm{I}_0 \exp(-\mu x) \tag{2.1}$$

where I_0 and I_t are the incident and transmitted beam intensities, respectively; x is the length of the X-ray path; and $\mu(x)$ is the X-ray linear attenuation coefficient, which varies with tissue type and hence is a function of the distance x through the patient. The integral of the attenuation coefficient is from 0 to L.

The reconstruction algorithm requires measurements of this integral along many paths in the fan beam at each of many angles about the isocenter. The value of L is known, and I_0 is determined by a system calibration. Hence values of the integral along each path can be determined from measurements of I_t.

2.4.3 X-Ray Detectors

X-ray detectors used in CT systems must;

a. have a high overall efficiency to minimize the patient radiation dose
b. have a large dynamic range
c. be very stable with time
d. be insensitive to temperature variations within the gantry.

These important factors contributing to the detector efficiency are geometric efficiency, quantum (also called *capture*) efficiency, and conversion efficiency.

The image receptors that are utilized in CT are referred to as detectors. The CT process essentially relies on collecting attenuated photon energy and converting it to an electrical signal. This analog signal will then be converted to a digital signal for computer reconstruction. A detector is a crystal or ionizing gas that when struck by an X-ray photon produces light or electrical energy. The two main types of detectors utilized in CT systems are scintillation or solid-state and xenon gas detectors. Saying simply, scintillation detector is a crystal that fluoresces under the action of an incident X-ray photon and produces light energy. The whole construction consists of a photodiode which is attached to the scintillation portion of the detector. The first one, the photodiode, transforms the light energy into electrical or analog energy. The strength of the detector signal is proportional to the number of attenuated photons that are successfully converted to light energy and then to an electrical or analog signal. The most frequently used scintillation crystals are made of bismuth germinate ($Bi_4Ge_30_{12}$) and cadmium tungstate ($CdWO_4$). Earlier designs utilized sodium and cesium iodide as the light-producing agent. One of the problems associated with these crystals were that at times it would fluoresce more than necessary. The afterglow problems associated with sodium and cesium iodide altered the strength of the detector signal which could cause inaccuracies during computer reconstruction.

> *Geometric efficiency* refers to the area of the detectors sensitive to radiation as a fraction of the total exposed area. Thin septa between detector elements to remove scattered radiation, or other insensitive regions, will degrade this value.
> *Quantum efficiency* refers to the fraction of incident X-rays on the detector that are absorbed and contribute to the measured signal.
> *Conversion efficiency* refers to the ability to accurately convert the absorbed X-ray signal into an electrical signal. It is not the same as the energy conversion efficiency.
> *Overall efficiency* is the product of the three, and it generally lies between 0.45 and 0.85. A value of less than 1 indicates a nonideal detector system and results in a required increase in patient radiation dose if image quality is to be maintained.
> The term *dose efficiency* sometimes has been used to indicate overall efficiency.

Modern commercial systems use one of two detector types: solid-state and gas ionization detectors.

1. *Solid-state detectors*

   Solid-state detectors consist of an array of scintillating crystals and photodiodes, as illustrated in Figure 2.3. The scintillators generally are either cadmium tungstate ($CdWO_4$) or a ceramic material made of rare earth oxides, although previous scanners have used bismuth germinate crystals with photomultiplier tubes. Solid-state detectors generally have very high quantum and conversion efficiencies and a large dynamic range.

2. *Gas ionization detectors*

   Gas ionization detectors, as illustrated in Figure 2.3, consist of an array of chambers containing compressed gas (usually xenon at up to 30 atm pressure). A high voltage is applied to tungsten septa between chambers to collect ions produced by the radiation. These detectors have excellent stability and a large dynamic range; however, they generally have a lower quantum efficiency than solid-state detectors.

In medical X-ray imaging, the conventional image intensifiers are being replaced step by step by digital flat panel (FD) technology. For example, there are FD-based X-ray systems for mammography or stationary *C* arm systems for angiography and cardiology. Besides already commercially available products, there are several preclinical studies with mobile or stationary prototype systems equipped with FD.

Not only obvious advantages on the basis of the geometric properties such as compact design, large field of view e.g., $40 \times 30\ \text{cm}^2$, and the resulting excellent patient access have led to this development but also the physical properties. Among the latter are extended dynamic range, up to 17 bit, high speed ~60 fps, and low-noise readout electronics, independence of the terrestrial magnetic field and a long life cycle with respect to inherent radiation damage.

However, the variable properties offset, gain, lag for each pixel of the detector matrix caused by the panel design-combination of photodiode, thin film transistor, line/column driver, amplifier, sample and hold unit, A/D converter have to be corrected for every single detector mode, e.g., field of view, binning, frame rate to bring out the best in image quality.

A comparison chart of conventional X-ray radiography and CT is given in Table 2.1.

Table 2.1 X-Ray Versus CT Scan

Imaging Modality	X-Ray	CT Scan
	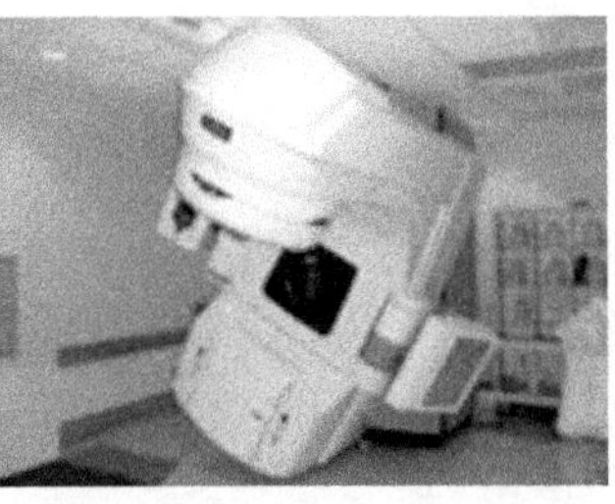	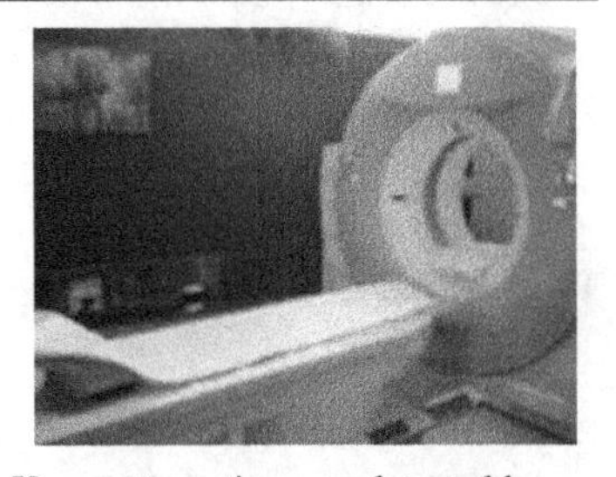
Principle	X-rays attenuated by denser tissue creates a shadow on the image.	X-ray attenuation was detected by detector and DAS system, follow by mathematical model to calculate the value of pixelism then become an image.
Acronym	X-radiation or Rontgen radiation.	Computed (axial) tomography.

(*Continued*)

Table 2.1 (Continued)

Imaging Modality	X-Ray	CT Scan
Scope of application	X-ray is limited to examining a few body conditions only.	CT scan outline bone inside the body very accurately.
Image specifics	Demonstrates the difference between bone density and soft tissue.	Good soft tissue differentiation especially with intravenous contrast. Higher imaging resolution and less motion artifact due to fast imaging speed.
Details of soft tissues	None—only bone and other dense tissue can be seen.	A major advantage of CT is that it is able to image bone, soft tissue, and blood vessels all at the same time.
Image specifics	Demonstrates the difference between bone density and soft tissue.	A major advantage of CT is that it is able to image bone, soft tissue, and blood vessels all at the same time.
What they show	X-rays are used to see metal objects inside the body that cannot be removed, e.g., metal plates, pace makers, and shrapnel	
Applications	X-ray technology is used to employ radiography (and other techniques) for diagnostic imaging. X-rays are useful to detect pathology of the skeletal system diseases and to detect certain diseases in the soft tissue (e.g., identification of pneumonia, pulmonary edema, lung cancer or abdominal, X-ray are helpful in detecting gallstones or kidney stones).	Suited for bone injuries, lung and chest imaging, cancer detection. Widely used in emergency room patient.
Ability to change the imaging plane without moving the patient	Does not have this ability.	With capability of Multidetector CT (MDCT), isotropic imaging is possible. After helical scan with multiplanar reformation function, an operator can construct any plane.
Time taken for complete scan	A few seconds.	Usually completed within 5 min. Actual scan time is usually less than 30 s.
Radiation exposure	Exposure to dangerous ionizing radiation.	The effective radiation dose from CT ranges from 2 to 10 mSv, which is about the same as the average person receives from background radiation in 3–5 years. Usually, CT is not recommended for pregnant women or children unless absolutely necessary.

(*Continued*)

Table 2.1 (Continued)

Imaging Modality	X-Ray	CT Scan
Details of bony structures	Detailed images of bone structure on photographic film as bones absorb X-rays, and X-rays affect photographic film in the same way as light.	Provides good details about bony structures.
Effects on the body	Provides good details about bony structures and can also alter the DNA.	Despite being small, CT can pose the risk of irradiation. Painless, noninvasive.
Cost	X-ray is relatively cheaper than CT scan.	CT scan costs range from $1200 to $3200—they are usually more expensive than X-ray.
Advantages and disadvantages	X-rays are used to treat malign tumors before its spreads throughout the human body. They help radiologists identify cracks, infections, injury, and abnormal bones. They also help in identifying bone cancer. X-rays help in locating alien objects inside the bones or around them. X-rays makes our blood cells to have higher level of hydrogen peroxide which could cause cell damage. A higher risk of getting cancer from X-rays. The X-rays are able to change the base of the DNA causing a mutation.	Due to the high contrast resolution of CT scanning differences between the tissues are more apparent compared to other techniques. Further, imaging in CT scan removes the superimposition of the structures. The data from other than the area of interest. The data from a single procedure can be viewed in different plans and thus increases diagnostic ability. This technique is also popular because it can be used to diagnose a number of conditions. This might eliminate the need of other diagnostic techniques such as colonoscopy. CT scan is associated with the risk of causing cancers such as lung cancer, colon cancer, and leukemia. This is mainly due to the use of X-rays. There are other safety concerns associated with the use of contrast agents which are administered intravenously. However, these disadvantages can be reduced with the use of lower doses.

References

[1] A.M. Cormack, Reconstruction of densities from their projections, with applications in radiological physics, Phys. Med. Biol. 18 (2) (1973) 195.

[2] G.N. Hounsfield, Computerized transverse axial scanning (tomography)—Part 1. Description of the system, Br. J. Radiol. 46 (1973) 1016.

[3] A.M. Cormack, Nobel Lecture, 8 December, 1979. Nobelprize.org.
[4] A.R. Margullus, J.H. Sunshine, Radiology at the turn of millennium, Radiology (2000).
[5] J.H. Siewerdsen, D.A. Jaffray, Cone-beam computed tomography with a flat-panel imager, Med. Phys. 28 (2001) 2.
[6] J.H. Siewerdsenm, O. Malley, Computed tomography, IBBME.
[7] O. Linton, Moments in Radiology History. Part 1–Part 7. Copyright © 2012 AuntMinnie.com.
[8] R. Baba, K. Ueda, M. Okabe, Using a flat-panel detector in high resolution cone beam CT for dental imaging, Dentomaxillofacial Radiol. 33 (2004) 285.
[9] Chicago Tribune Business, February 18 (1988).

3 Physics of Magnetic Resonance Imaging

3.1 History of Magnetic Resonance Imaging in Brief

Briefly, magnetic resonance imaging (MRI) is based on a physical phenomenon called nuclear magnetic resonance or NMR, in which magnetic fields and radio waves cause atoms to give off tiny radio signals. The discovery of this phenomenon dates to 1930. Felix Bloch from Stanford University and Edward Purcell from Harvard University discovered NMR. In his Nobel Prize winning paper, Felix Bloch proposed rather new properties for the atomic nucleus. As he stated, the nucleus behaves like a magnet. A charged particle, such as a proton, spinning around its own axis has a magnetic field, known as a magnetic momentum. Bloch wrote down his finding in what we know as the Bloch Equations. It would take until the early 1950s before his theories could be verified experimentally. In 1960, NMR spectrometers were introduced for analytical purposes. NMR spectroscopy was then used as means to study the composition of chemical compounds [1–4].

The 2003 Nobel Prize in Physiology and Medicine was awarded to Paul C Lauterbur and Peter Mansfield for their discoveries concerning MRI.

Lauterbur, a professor of Chemistry at the State University of New York at Stony Brook wrote a paper on a new imaging technique. He termed this imaging modality zeugmatography (from the Greek "zeugmo" meaning yoke or a joining together). In 1975, Richard Ernst described the use of Fourier transform of phase and frequency encoding to reconstruct 2D images. This technique is the basis of today's MRI. In April 1974, Lauterbur gave a talk at a conference in Raleigh, North Carolina. Ernst attended this conference and realized that instead of Lauterbur's back-projection, one could use switched magnetic field gradients in the time domain. This led to the 1975 publication that described for the first time a practical method to rapidly reconstruct an image from NMR signals. Ernst was rewarded for his achievements in pulsed Fourier transform NMR and MRI with the 1991 Nobel Prize in Chemistry. MRI gives an ability to look inside the body without dissection (Figure 3.1).

The early contribution of computed tomography (CT) to MRI is worth noting. Hounsfield introduced X-ray-based CT in 1973 the same year that Lauterbur and Mansfield introduced spatial localization of NMR signals to produce 2D images. This date is important to the MRI timeline. It demonstrated the interest of the scientific and clinical communities in noninvasive cross-sectional *in vivo* imaging.

Medical Imaging Technology. DOI: http://dx.doi.org/10.1016/B978-0-12-417021-6.00003-4

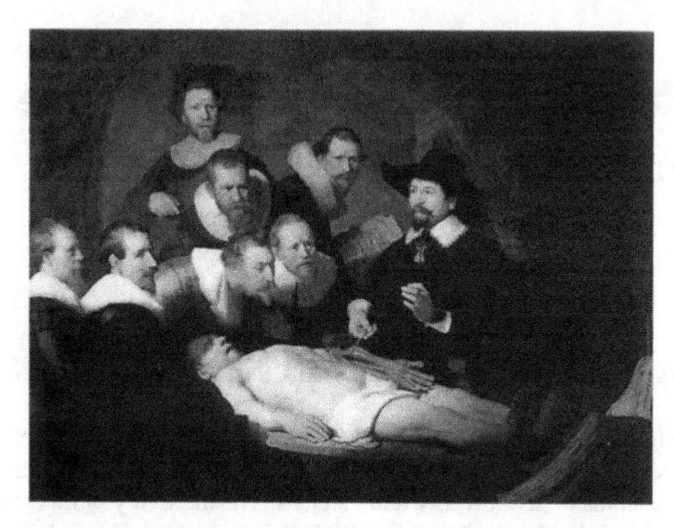

Figure 3.1 The anatomy lesson of Dr. Nicolaes Tulp by Rembrandt van Rijin.

Lauterbur imaging experiments moved science from the single dimension of NMR spectroscopy to the second dimension of spatial orientation—the foundation of MRI.

Mansfield of Nottingham, England, further developed the utilization of gradients in the magnetic field. He showed how the signals could be mathematically analyzed, which made it possible to develop a useful imaging technique. Mansfield also showed how extremely fast imaging could be achievable. This became technically possible within medicine a decade later.

In 1970, Raymond Damadian, a medical doctor and research scientist, discovered the basis for using MRI as a tool for medical diagnosis. He found that different kinds of animal tissue emit response signals that vary in length and that cancerous tissue emits response signals that last much longer than noncancerous tissue.

Less than 2 years later he filed his idea for using MRI as a tool for medical diagnosis with the US Patent Office, entitled "Apparatus and Method for Detecting Cancer in Tissue." A patent was granted in 1974, it was the world's first patent issued in the field of MRI. By 1977, Damadian completed construction of the first whole-body MRI scanner, which he dubbed the "Indomitable."

The medical use of MRI is very impressive and developed rapidly. The first MRI equipment in health was available at the beginning of the 1980s. In 2002, approximately 22,000 MRI cameras were in use worldwide and more than 60 million MRI examinations were performed.

One could discuss who was responsible for bringing MRI to us, although, in all fairness, one could accept that both gentlemen had their contribution.

The name Nuclear Magnetic Resonance (NMR) was changed into Magnetic Resonance Imaging (MRI) because it was recognized that the word nuclear would not find wide acceptance among the public.

3.2 Introduction to MRI Physics and Basic Concepts

Two main questions a physicist working in MRI are frequently asked by medical doctors and physicians sound as follows:

1. What is MRI?
2. How does it work?

The physicists usually answer in rhetorical manner: How much Physics do you know? Moreover, how far are you prepared to go?

Like everything else in our life, MRI can be explained at different levels, advancing more and more as the physical knowledge progresses.

To understand MRI, we first need to know basic physics at the molecular, atomic, and subatomic level.

All matter is composed of molecules. These latter, in turn, are composed of atoms. By accepted definition, atom is the smallest unit of matter that is unique to a particular chemical element. It consists of protons, neutrons, and electrons. The core or nucleus of an atom consists of protons and neutrons. Electrons move around the nucleus. The protons carry a positive charge; while electrons carry a negative charge (neutrons do not have a charge).

The starting point is the production of magnetic field. Let us assume that electron travels or moves along a wire. As a result, a magnetic field is produced around, as shown in Figure 3.2.

Suppose now that electric current flows in a wire. In such a case, loop is formed and a large magnetic field will be formed perpendicular to the loop (Figure 3.2).

In physics, the resonance means the increase in amplitude of oscillation of an electric or mechanical system exposed to a periodic force whose frequency is equal or very close to the natural undamped frequency of the system and aids an efficient transfer of energy [4].

In MR system, the main magnetic field is produced by a large electric current flowing through wires that are formed into a loop in the magnet of the imaging system (Figure 3.3). A typical clinical MR system will have a magnetic field strength of 1.5 T (Tesla) (1 T = 10.000 Gauss). The wires are immersed in liquid helium

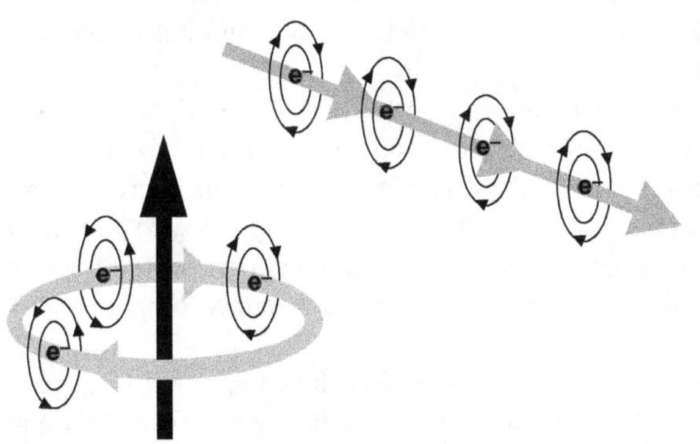

Figure 3.2 Electron moves around a wire.

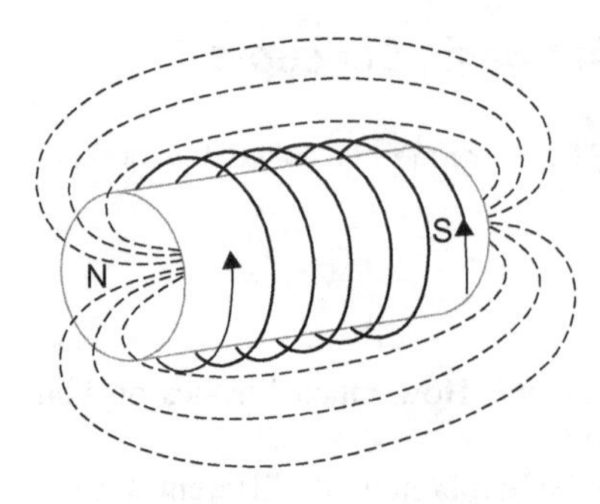

Figure 3.3 Main magnetic field used in MRI systems. A large electric current in loops of wire at superconducting temperatures will produce a very large magnetic field. N—North, S—South.

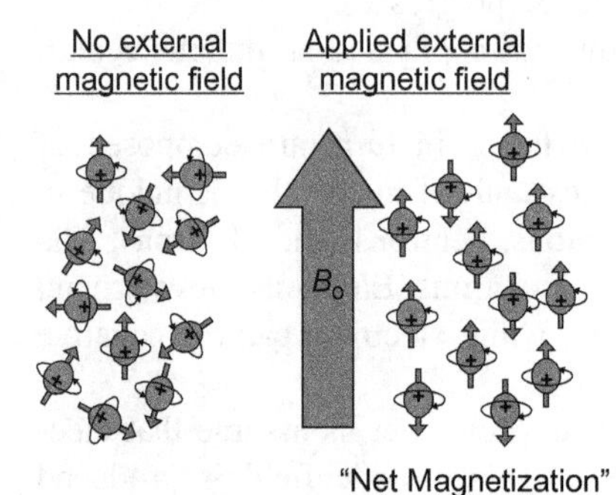

Figure 3.4 If there is no external magnetic field, hydrogen protons (+) is oriented randomly. At the same time, when the protons are placed in a strong magnetic field (B_0), a net magnetization will be produced parallel to the main magnetic field. In this case, one can observe alignment of protons with the B_0 field.

(at superconducting temperatures) so one can use very large currents to produce the strong magnetic field.

To inject electric current into the coils of wire, the magnet can be "ramped" with a power supply and the power supply can then be removed. The imaging system can retain this electric current for many years. There is no need to inject additional electric current. If these conditions are fulfilled, one can reach minimal losses in electric current and minimal decrease in magnetic field strength. The liquid helium levels in the magnet will need to be filled at regular intervals (once per month to once every few years, depending on the magnet design) [5–8].

Putting these basic elements together, there are protons in the body, positively charged and spinning about their axes that act like tiny magnets. Here we note that they are randomly oriented so that their magnetic fields do not sum but rather cancel out (Figure 3.4).

When we place these protons in a strong magnetic field (called B_0), some will tend to align in the direction of the magnetic field and some will tend to align in a direction opposite to the magnetic field. The magnetic fields from many protons will cancel out, but a slight excess of the protons will be aligned with the main magnetic field, producing a "net magnetization" that is aligned parallel to the main magnetic field. This net magnetization becomes the source of our MR signal and is used to produce MR images.

It is necessary to introduce a reference to a direction. It is important to discuss the coordinate system, which will orient us for future discussion. The direction

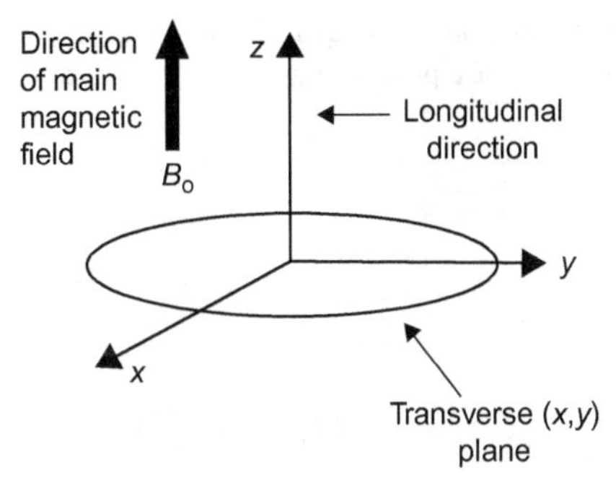

Figure 3.5 Coordinate system (usually exploited). For a typical 1.5-T cylindrical-bore imaging unit, the z-axis (longitudinal direction) is often aligned with the main magnetic field. On the opposite side, the plane perpendicular to this is called the transverse plane.

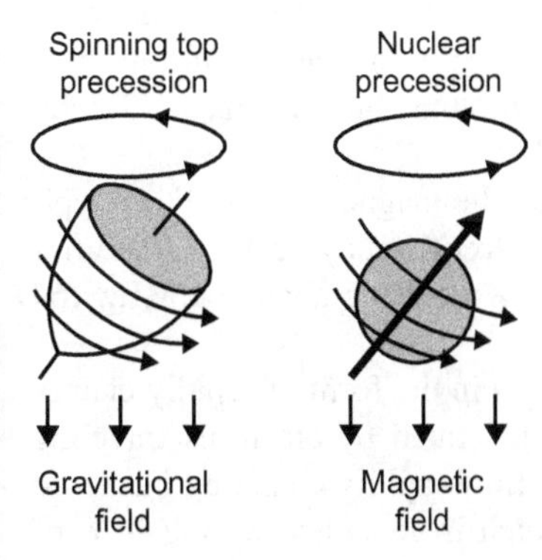

Figure 3.6 Schematic illustration of precession. Note that precession of a spinning top and nuclear precession are similar in that an external force combined with the spinning motion causes precession.

parallel to the main magnetic field is the longitudinal direction, which may also be called the z-direction (Figure 3.5).

For typical 1.5-T superconducting cylindrical-bore magnets, the z-direction is horizontal and corresponds to the head-to-foot (or foot-to-head) direction. The plane perpendicular to this direction is called the transverse plane or the x−y plane. For a patient who is headfirst and supine in a superconducting magnet, the x-direction is often chosen to be the left-right direction of the patient and the y-direction is often chosen to be the anterior-posterior direction. Interestingly, the transverse plane matches the axial plane for typical 1.5-T magnets.

A spinning top spins about its axis. The force of gravity attempts to pull the top in a way that it will fall down. The combined effects of gravity, from one side, and the spinning motion, from the other side, cause the top to precess. The same thing happens with nuclear precession. There are protons that are spinning and acting like tiny magnets. If we place these spinning protons in a strong magnetic field, the force from the magnetic field interacts with the spinning protons and results in precession of the protons (Figure 3.6).

That is important here is the frequency of precession. In other words, how many revolutions in a second does the proton precess? We must know this precessional frequency as we must know the frequency of the pendulum motion. It is this proton precessional frequency that allows us to create a situation through which the resonance phenomenon can be used to efficiently transfer energy to the protons.

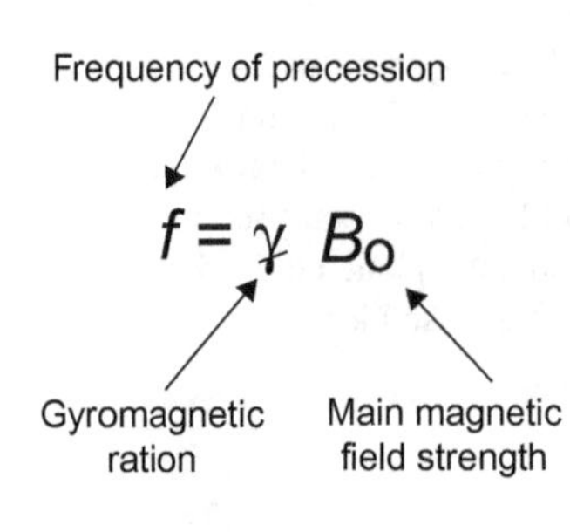

Figure 3.7 Larmor equation. This equation allows us the frequency of precession of a proton in a magnetic field to be determined.

The proton precessional frequency is determined from the Larmor equation, in which the frequency of precession, *f*, is equal to a constant times the main magnetic field strength (Figure 3.7). The constant is called the gyromagnetic ratio and is a characteristic of each type of nuclei. For hydrogen protons, the gyromagnetic ratio is equal to 42.6 MHz/T (megahertz per Tesla) [6,7].

The main magnetic field strength, B_0, depends on the magnet design. For a typical superconducting MR system, the magnetic field strength may be 1.5 T. The frequency of precession then will equal 42.6 MHz/T × 1.5 T or about 64 MHz (64 million times per second).

In MR systems, radiofrequency (RF) energy comes in the form of rapidly changing magnetic and electric fields. These latter are generated by electrons traveling through loops of wire with the direction of current flow rapidly changing back and forth at "radiofrequencies." The magnetic field which is generated by the flow of electrons will also rapidly change directions. Radio and TV stations broadcast at frequencies in units of megahertz, so a broadcast at 89.9 on your FM dial is really at 89.9 MHz. This RF energy is close to the precessional frequencies of a 1.5-T magnet (64 MHz). Reasonably, MR systems must be shielded from external RF signals.

For the MR system, this RF energy is transmitted by an RF transmit coil (e.g., body coil, head coil, and knee coil). Typically, the RF is transmitted for a short period of time and referred to as an RF pulse. Notably that this transmitted RF pulse must be at the precessional frequency of the protons (determined by the Larmor equation) in order for resonance and, in addition, for efficient transfer of energy from the RF coil to the protons.

When protons in our body are placed in the vicinity of a strong magnetic field, the magnetic fields from these protons combine to form a net magnetization. This net magnetization points in a direction parallel to the main magnetic field. The direction is also called the longitudinal direction. As energy is absorbed from the RF pulse, the net magnetization rotates away from the longitudinal direction (Figure 3.8). The amount of rotation (termed the flip angle) depends on the strength and duration of the RF pulse [5–8].

For the case of the RF pulse causing the net magnetization to rotate in the transverse plane, it we say about a 90° RF pulse. In an opposite case, when the RF pulse rotates the net magnetization 180° into the z-direction, that is termed a 180° RF pulse. Important that one can control the net magnetization rotation to any angle by

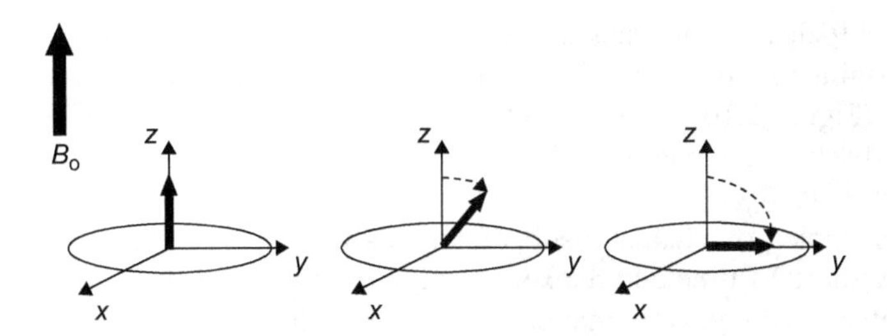

Figure 3.8 Illustration of how RF energy is absorbed. Prior to an RF pulse, the net magnetization shown by small black arrow is aligned parallel to the main magnetic field and the z-axis (at the left). An RF pulse at the Larmor frequency will allow energy to be absorbed by the protons and cause the net magnetization to rotate away from the z-axis (center and right).

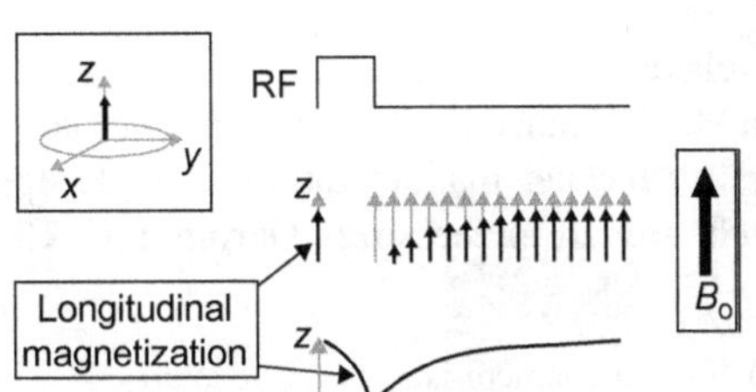

Figure 3.9 Schematics of longitudinal or T1 relaxation. If a 90° RF pulse is applied, longitudinal magnetization becomes 0. With time, the longitudinal magnetization grows back in a direction parallel to the main magnetic field.

the strength and/or duration of the RF pulse. The above mentioned 90° and 180° RF pulses seemed to be useful when we try to discuss the spin echo (SE) and fast imaging techniques.

3.3 T1 Relaxation

One may call longitudinal magnetization as net magnetization that is aligned in the longitudinal direction. After a 90° RF pulse rotates the longitudinal magnetization into the transverse plane, this magnetization may be recalled as transverse magnetization. After a 90° RF pulse, the longitudinal magnetization is 0. The magnetization then increases in the longitudinal direction (Figure 3.9).

This is called longitudinal relaxation or T1 relaxation.

The rate at which this longitudinal magnetization grows back is different for protons associated with different tissues and is the fundamental source of contrast in T1-weighted images.

T1 is a parameter characteristic of specific tissue. Undoubtedly, it depends on the main magnetic field strength and, at the same time, is related to the rate of regrowth of longitudinal magnetization.

The net magnetization does not rotate back up—it rather increases in a direction always parallel to the longitudinal direction (the direction of the main magnetic field).

Figure 3.9 clearly illustrates this effect. We can define T1 as the time necessary for the longitudinal magnetization to reach 63% of its final value, assuming a 90° RF pulse (Figure 3.10). The magnetization of tissues with different values of T1 will grow back in the longitudinal direction. Reasonably, these processes take place at different rates.

Few examples of characteristic of our body are given below. White matter has a very short T1 time and relaxes rapidly. Cerebrospinal fluid (CSF) has a long T1 and relaxes slowly. Gray matter has an intermediate T1 and relaxes at an intermediate rate (Figure 3.11). If we try to obtain an image at a time when these curves were widely separated, we need an image with high contrast between these tissues.

3.4 T2 Relaxation

Figure 3.12 exemplifies T2 (or transverse) relaxation. T2 begins with the net magnetization aligned with the z-direction and a 90° RF pulse that rotates this net magnetization into the transverse plane. Generally, the net magnetization is made up of contributions from many protons, which are all precessing. During the RF

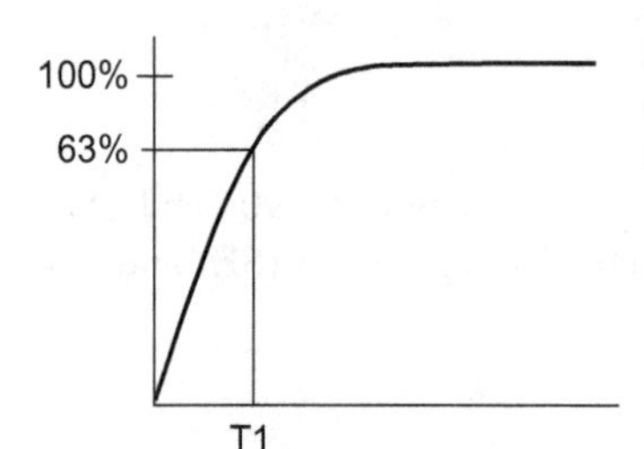

Figure 3.10 T1 is a characteristic of tissue and is defined as the time that it takes the longitudinal magnetization to grow back to 63% of its final value.

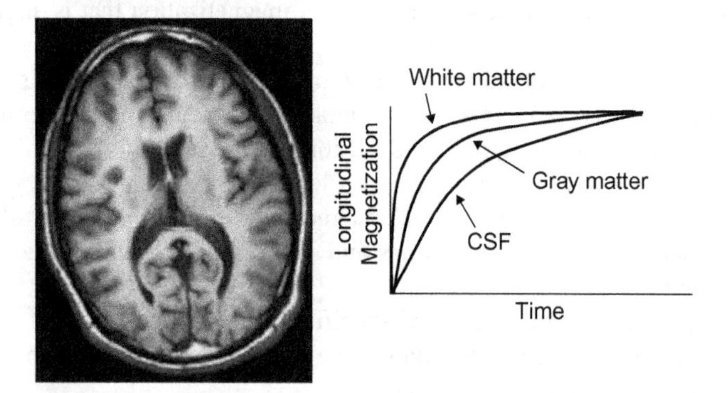

Figure 3.11 T1-weighted contrast. Different tissues have different rates of T1 relaxation. If an image is obtained at a time when the relaxation curves are clearly separated, T1-weighted contrast will be maximized. Magnetization (Mag) is shown at the ordinate axes.

pulse, the protons begin to precess together (they become "in phase"). Immediately after the 90° RF pulse, the protons are still in phase but begin to dephase due to several effects.

Among effects that cause T2 dephasing, spin−spin interaction, magnetic field inhomogeneity, magnetic susceptibility, and chemical shift effects should be mentioned.

The amount of transverse magnetization can be measured by a receiver coil. It should be mentioned that an electric current in a wire will produce a magnetic field perpendicular to the loop of wire. Measurement of the transverse magnetization (our "MR signal") occurs through an opposite effect. In this case, the transverse magnetization, which is a magnetic field, can induce a current in a loop of wire (Figure 3.13). This, in turn, induced electric current which is then digitized and recorded in the computer of the MR system for later reconstruction as MRI.

3.5 The Spin Echo

The SE can be used to recover dephasing due to all effects (except spin−spin interactions). After a 90° RF pulse, protons that were in phase begin to dephase in the transverse plane (Figure 3.14).

If 180° RF pulse is applied, the spins will rotate over to the opposite axis. Now, the spins will begin to rephase.

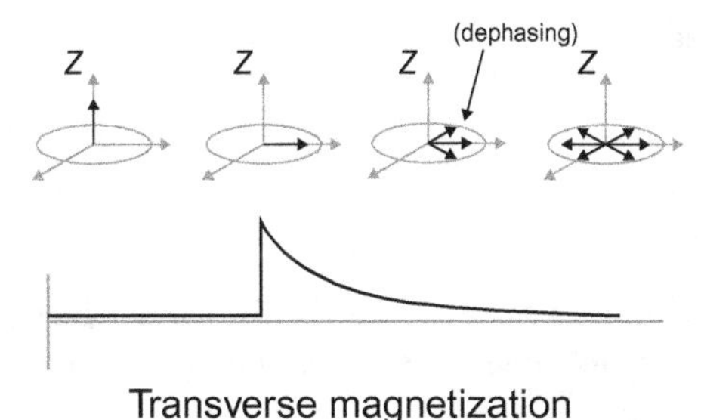

Figure 3.12 Transverse (T2*) relaxation. After application of a 90° RF pulse, transverse magnetization is maximized; it then begins to diphase. The signals from these dephasing protons begin to cancel out. As a result, the MR signal decrease.

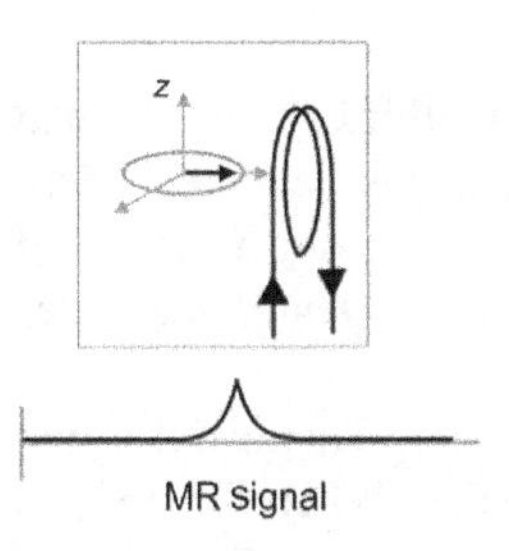

Figure 3.13 Measurements of MR signal (see the text for details).

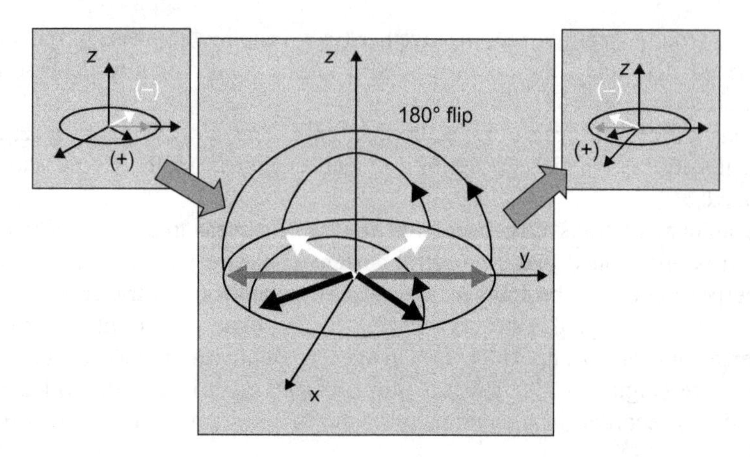

Figure 3.14 Mechanism of SE.

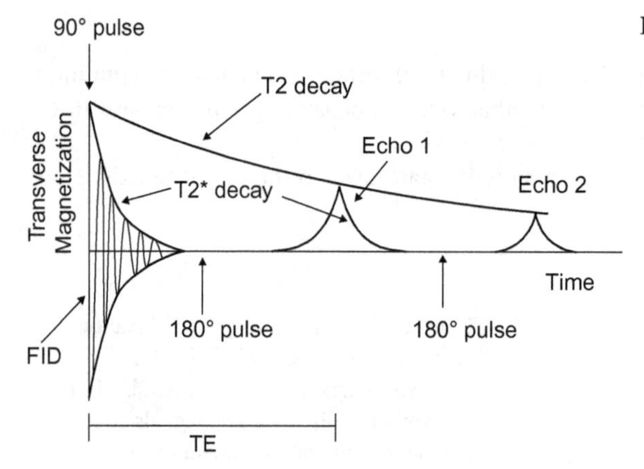

Figure 3.15 Formation of SEs.

The spins will come back together and the signal measured with our receiver coil will increase, from a maximum signal, and then decrease as the spins once again dephase (Figure 3.15).

3.6 MRI Source

Our body consists of 80% water. Water constitutes about two-thirds of the human body weight, and this high water content explains why MRI has become widely applicable to medicine. There are differences in water content among tissues and organs. In many diseases, the pathological process results in changes of the water content, and this is reflected in the MRI.

Water is a molecule composed of hydrogen and oxygen atoms. The nuclei of the hydrogen atoms are able to act as microscopic compass needles. When the body is exposed to a strong magnetic field, the nuclei of the hydrogen atoms are directed into order—stand "at attention." When submitted to pulses of radio waves, the energy content of the nuclei changes. After the pulse, a resonance wave is emitted when the nuclei return to their previous state.

From our chemistry lessons, we know that there are many different elements, 110 to be precise. Because we exist mainly of water let us have a look at it. Water consists of 2 hydrogen and 1 oxygen atom. The hydrogen atom (the first element in the periodic table) has a nucleus, called proton, and 1 moon, called electron.

This proton is electrically charged and it rotates around its axis. There we have the analogy with the earth. Also the hydrogen proton can be looked at as if it were a tiny bar magnet with a north and a south pole.

Let us take hydrogen as the MR the dominant imaging source (Figure 3.16). There are two reasons for this. Firstly, all we have a lot of them in our body. Actually it is the most abundant element we have. Secondly, in quantum physics there is a thing called "gyromagnetic ratio." It is beyond the scope of this story what it represents; suffice to know that this ratio is different for each proton. It just so happens, that this gyromagnetic ratio for hydrogen is the largest; 42.57 MHz/T.

Hydrogen is not the only element we can use for MRI. In fact any element, which has an odd number of particles in the nucleus, can be used. Some elements, which can be used, are shown in Table 3.1 [4,5]. A comparison chart of CT scan and MRI is given in Table 3.2.

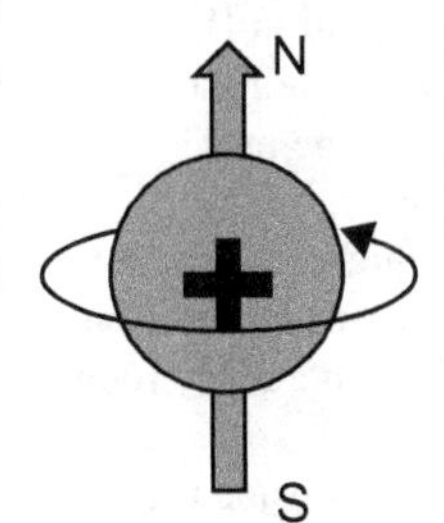

Figure 3.16 Hydrogen proton. The hydrogen proton is positively charged and spins about its axis. This positively charged spinning proton acts like a tiny magnet. The hydrogen protons in our body thus act like many tiny magnets.

Table 3.1 MRI "Suitable" Elements

Isotope	**Symbol**
Hydrogen	^{1}H
Carbon	^{13}C
Oxygen	^{17}O
Fluorine	^{19}F
Sodium	^{23}Na
Magnesium	^{25}Mg
Phosphorus	^{31}P
Sulfur	^{33}S
Iron	^{57}Fe

Table 3.2 CT Scan Versus MRI

Imaging Technique	MRI	CT Scan
Acronym	Magnetic resonance imaging	Computed (axial) Tomography

Principle	Body tissue that contains hydrogen atoms (e.g., in water) is made to emit a radio signal which is detected by the scanner.	X-ray attenuation was detected by detector and DAS system, follow by mathematical model (back-projection model) to calculate the value of pixelism then become an image.
History	First commercial MRI in 1981, with significant increase in MRI resolution and choice of imaging sequences over time.	The first commercially viable CT scanner was invented by Sir Godfrey Hounsfield in Hayes, United Kingdom; the first patient brain-scan was done on 1 October 1971.
Application	Suited for soft tissue evaluation, e.g., ligament and tendon injury, spinal cord injury, and brain tumors.	Suited for bone injuries, lung and chest imaging, cancer detection. Widely used in emergency room patient.
Scope of application	MRI is more versatile than the X-ray and is used to examine a large variety of medical conditions.	CT can outline bone inside the body very accurately.
Image specifics	Demonstrates subtle differences between the different kinds of soft tissues.	Good soft tissue differentiation especially with intravenous contrast. Higher imaging resolution and less motion artifact due to fast imaging speed.
Principle used for imaging	Uses large external field, RF pulse, and three different gradient fields.	Uses X-rays for imaging.
Details of bony structures	Less detailed compared to X-ray.	Provides good details about bony structures.

(*Continued*)

Table 3.2 (Continued)

Imaging Technique	MRI	CT Scan
Details of soft tissues	Much higher soft tissue detail as compared to CT scan.	A major advantage of CT is that it is able to image bone, soft tissue, and blood vessels all at the same time.
Time taken for complete scan	Scanning typically run for about 30 min.	Usually completed within 5 min. Actual scan time usually less than 30 seconds. Therefore, CT is less sensitive to patient movement than MRI.
Cost	MRI costs range from $1200 to $4000 (with contrast); which is usually more than CT scans and X-rays, and most examining methods.	CT scan costs range from $1200 to $3200; they usually cost less than MRIs (about half the price of MRI).
Ability to change the imaging plane without moving the patient	MRI machines can produce images in any plane. Plus, 3D isotropic imaging also can also produce multiplanar reformation.	With capability of MDCT, isotropic imaging is possible. After helical scan with multiplanar reformation function, an operator can construct any plane.
Cost	MRI costs range from $1200 to $4000 (with contrast); which is usually more than CT scans and X-rays, and most examining methods.	CT scan costs range from $1200 to $3200; they usually cost less than MRIs (about half the price of MRI).
Radiation exposure	None. MRI machine control/limit energy deposition in patient.	The effective radiation dose from CT ranges from 2 to 10 mSv, which is about the same as the average person receives from background radiation in 3–5 years. Usually, CT is not recommended for pregnant women or children unless absolutely necessary.
Effects on the body	No biological hazards have been reported with the use of the MRI.	Despite being small, CT can pose the risk of irradiation. Painless, noninvasive.
Limitation for scanning patience	Patients with cardiac pacemakers are not allowed to get MRI scan, tattoos and metal implants may be contraindicated due to possible injury to patient or image distortion (artifact). Patient over 350 lb may be over table weight limit.	Patients with any metal implants can get CT scan. A person who is very large (e.g., over 450 lb) may not fit into the opening of a conventional CT scanner or may be over the weight limit for the moving table.

(*Continued*)

Table 3.2 (Continued)

Imaging Technique	MRI	CT Scan
Intravenous contrast agent	Very rare allergic reaction. Risk of nephrogenic systemic fibrosis with free gadolinium in the blood and severe renal failure. It is contraindicated in patients with GFR under 60 and especially under 30 ml/min.	Nonionic iodinated agent is covalently bind the iodine and have fewer side effects. Allergic reaction is rare but more common than MRI contrast. Risk of contrast induced nephropathy (especially in renal insufficiency (GFR $<$ 60), diabetes and dehydration.

References

[1] O. Linton, History of MRI, Soc. Radiol. JSR (2007).
[2] A Short History of MRI, Tesla Memorial Society, New York. Available from: www.teslasociety.com.
[3] A.Q. Prasad, The anatomy of MRI invention, Publ. Med. 37 (2007) 533.
[4] A. Filler, The history of CT, MRI, The Internet Journal of Neurosurgery 7 (2010). Available from: http://dx.doi.org/10.1038/nprc2009.
[5] R. Pooley, Fundamental physics of MRI imaging, Radio Graph. 25 (2005) 1087.
[6] M. Forja, Basic Principles of MRI (2012).
[7] E.W. Blink, MRI, Physics (2004).
[8] M Gray, Magnetic resonance imaging, Radiographics 89 (2007) 4.
[9] A Song, Principles of MRI, Duke University, 2006.

4 Positron Emission Tomography

4.1 Introduction

By the definition, positron emission tomography, also referred to as PET imaging or a PET scan, is a type of nuclear medicine imaging. PET is a nuclear medicine technique that produces 3D images (pictures) of metabolic processes in the body and allows visualizing the body at the cellular and functional levels. The system detects pairs of gamma rays emitted indirectly by a positron-emitting radionuclide (tracer), which is introduced into the body on a biologically active molecule [1].

According to Med.Dictionary, PET is an imaging technique that uses radioactive substances to obtain 3D color images of human body functionalization. Short-lived radioactive substances are used. These images are called PET scans and the technique is termed PET scanning.

PET scanning provides information about the body's chemistry not available through other procedures. Unlike CT (computerized tomography) or MRI (magnetic resonance imaging), techniques that look at anatomy or body form, PET studies metabolic activity or body function and disease processes. Primarily, PET has been used in cardiology, neurology, and oncology.

Indeed, PET as a whole could not become a clinical specialty until the "typical" medical facility saw the need to offer PET as part of their diagnostic procedure capability.

Nuclear medicine is a branch of medical imaging that uses small amounts of radioactive material to diagnose and determine the severity or treat a variety of diseases, including many types of cancers, heart disease, gastrointestinal, endocrine, neurological disorders, and other abnormalities within the body. Because nuclear medicine procedures are able to pinpoint molecular activity within the body, they offer the potential to identify disease in its earliest stages as well as a patient's immediate response to therapeutic interventions.

Nuclear medicine imaging procedures are noninvasive and, with the exception of intravenous injections, are usually painless medical tests that help physicians diagnose and evaluate medical conditions. These imaging scans use radioactive materials called radiopharmaceuticals or radiotracers.

Depending on the type of nuclear medicine examination, the radiotracer is either injected into the body, swallowed, or inhaled as a gas and eventually accumulates in the organ or area of the body being examined. Radioactive emissions from the radiotracer are detected by a special camera or imaging device that produces pictures and detailed molecular information.

Medical Imaging Technology. DOI: http://dx.doi.org/10.1016/B978-0-12-417021-6.00004-6

In the last decade, the clinical value of PET as an imaging modality has become increasingly apparent. Medical professionals in the fields of oncology, cardiology, and neurology have been using PET techniques to assess metabolism in their respective evaluations of cancer, damaged heart tissue, and brain disorders. Expectations are high that PET, a nuclear medicine scanning procedure that employs positron-emitting radioactive isotopes to image the body's metabolic activity, will add a new dimension to the evaluation and treatment planning of a variety of diseases and medical conditions and that it will serve as a valuable tool for patients' follow-up and care.

4.2 Historical Sketch

The potential of positron imaging and the value of eliminating the collimator were recognized by the early developers of nuclear medicine instrumentation, long before the advent of reconstruction algorithms which could allow the generation of transverse sections from data covering a large number of angles [1–3].

One of the factors most responsible for the acceptance of positron imaging was the development of radiopharmaceuticals. In particular, the development of labeled 2-fluorodeoxy-D-glucose (2FDG) by the Brookhaven group under the direction of Al Wolf and Joanna Fowler was a major factor in expanding the scope of PET imaging.

Brownell and his group at the Massachusetts General Hospital contributed significantly to the development of PET technology and included the first demonstration of annihilation radiation for medical imaging [1]. Their innovations, including the use of light pipes and volumetric analysis, have been important in the deployment of PET imaging. In 1961, James Robertson and his associates at Brookhaven National Laboratory built the first single-plane PET scan nicknamed the "head-shrinker."

The compound was first administered to two normal human volunteers by Abass Alavi in August 1976 at the University of Pennsylvania. Brain images obtained with an ordinary (non-PET) nuclear scanner demonstrated the concentration of FDG in that organ. Later, the substance was used in dedicated positron tomographic scanners, to yield the modern procedure.

The logical extension of positron instrumentation was a design using two 2D arrays. PC-I was the first instrument using this concept and was designed in 1968, completed in 1969 and reported in 1972. The first applications of PC-I in tomographic mode as distinguished from the computed tomographic mode were reported in 1970. It soon became clear to many of those involved in PET development that a circular or cylindrical array of detectors was the logical next step in PET instrumentation. Although many investigators took this approach, James Robertson and Z.H. Cho were the first to propose a ring system that has become the prototype of the current shape of PET.

Hal Anger also worked on positron imaging almost as soon as he developed the single photon Anger camera. His dual detector PET scanner was made commercially available for a brief period by Nuclear Chicago Corp., but early clinical data quickly pointed out the main problem with PET imaging: while the elimination of the collimator increased the photon flux hitting the detectors by more than an order of magnitude relative to single photon imaging, the fraction of event which are found in coincidence is of the order of 1%, therefore requiring a high singles count rate capability to achieve an acceptable coincidence rate.

The early 1950s heralded two inventions and investigators that contributed to the emergence of imaging in nuclear medicine: the rectilinear scanner by Benedict Cassen and colleagues at UCLA and the gamma camera by Hal Anger and colleagues in Berkeley, CA.

In 1953, Gordon Brownell at MIT created a precursor to the up-and-coming PET scanner when he constructed the first detector device to record the annihilation that occurs when positrons from positron-emitting pharmaceuticals collide with electrons in the human body (Figure 4.1).

The success of prototype positron scanner led G. Brownell to develop scanner designed specifically for brain images Figure 4.2. The first computed image PET device dates to 1968–1971 (Figure 4.3).

In the middle 1950s, Michel Ter-Pogossian and William Powers of Washington University's Mallinckrodt Institute of Radiology reinstated in biomedical research the use of radiopharmaceuticals labeled with short-lived, cyclotron-produced radioisotopes. At about this time (1955) on the other side of the world, the first medical cyclotron was built at Hammersmith Hospital in London. Successful experimentation at Hammersmith Hospital led to installation of a National Institutes of Health (NIH)-funded cyclotron in Washington University Medical Center, followed by the Department of Energy's funding of hospital cyclotrons at UCLA, University of Chicago, and Memorial Sloan Kettering Institute in New York. And, so the production of radioisotopes continued with the use of cyclotrons at these hospitals as well as with the use of existing cyclotrons at UC Berkley and Ohio State. Another development that contributed to the emergence of PET was CT, which was

Figure 4.1 First clinical positron imaging device. Drs. Browner and Aronov are shown with the scanner (1953 [1]).

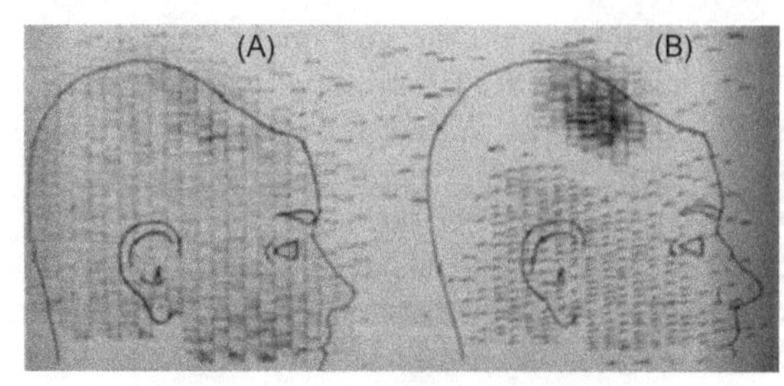

Figure 4.2 Coincidence and ambulance scans of patient with recurring brain tumor. Coincidence scan (A) of patient showing recurrence of tumor after under previous operation site, and unbalance scan (B) showing asymmetry to the left [1].

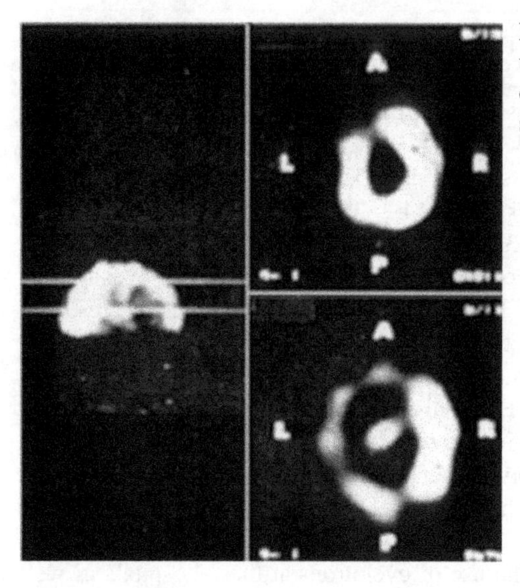

Figure 4.3 Images of a brain tumor study using PC1 device. A tumor is clearly observable in the lower transverse slice [1].

invented by Alan Cormack and Godfrey Hounsfield and popularized by investigator, David Kuhl, and colleagues at the University of Pennsylvania in 1959.

4.2.1 The 1960s and 1970s—The Development of PET

During the late 1960s, Ter-Pogossian and colleagues continued with and advanced their studies using radioisotope/radiopharmaceutical techniques while Kuhl and colleagues persevered in their studies of emission computed tomography (CT) and built the Mark II scanner, often referred to as the ancestor to today's CT scanners (Figure 4.4). Kuhl and colleagues also studied reconstruction.

4.3 Positrons

Positrons are the antiparticles of electrons. The major difference from electrons is their positive charge. Positrons are formed during decay of nuclides that have an excess of protons in their nucleus compared to the number of neutrons. When

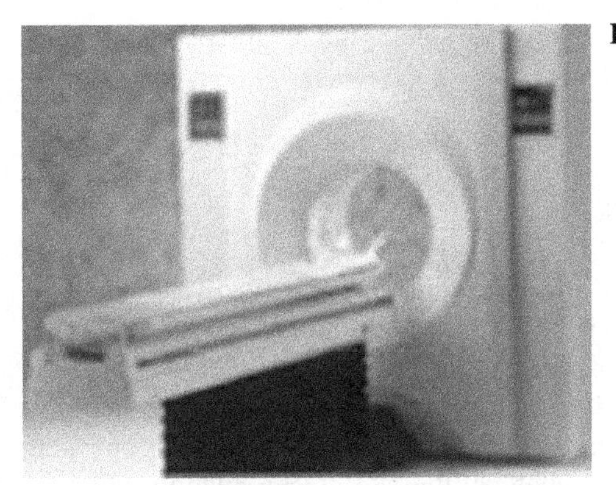

Figure 4.4 A PET scanner.

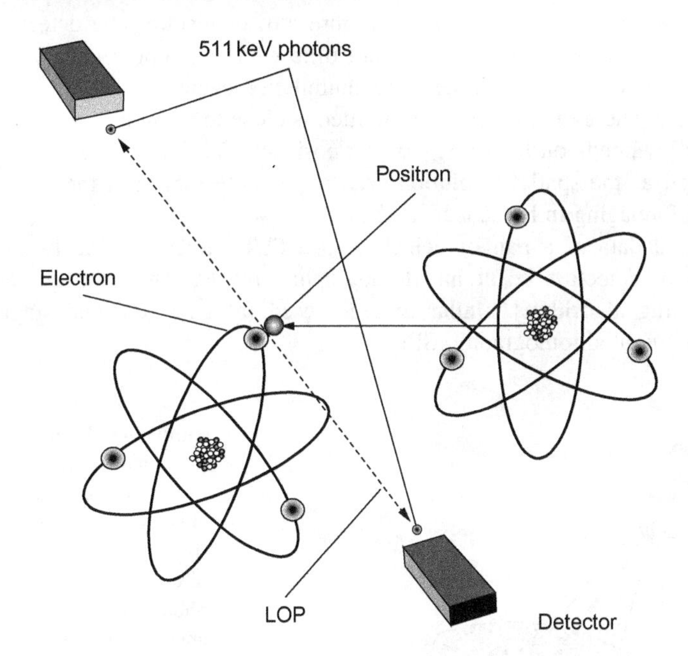

Figure 4.5 The positron electron annihilation system.

decaying takes place, these radionuclides emit a positron and a neutrino. While the neutrino escapes without interacting with the surrounding material, the positron interacts with an electron. During this annihilation process, the masses of the positron and the electron are converted into two photons that travel apart in almost opposite directions. Since the entire masses are being transformed, each photon obtains 511 keV of energy (Figure 4.5).

In the early days of PET, annihilation radiation was considered to be less optimal for imaging because of its relatively high energy. However, later it was demonstrated that the two annihilation photons travelling nearly collinearly offer substantial advantages in the collimation of this radiation. This property of the annihilation radiation was found particularly desirable in the 3D tomographic imaging of the distribution of positron emitters.

4.4 Detection of Positrons

Positrons cannot be detected directly—the maximum linear range of a positron is in the order of only a few millimeters. That is, in general, the positron cannot escape from the human body for external detection.

At the same time, the two annihilation photons can be detected (simultaneously) using two or more scintillation detectors in coincidence mode. When two opposing detectors are detecting the annihilation photons, the site of annihilation will be a point on the line of projection (LOP, Figure 4.6) connecting the detectors. If two photons are detected within a very small time window, approximately 15 ns, they are assumed to originate from the same annihilation event.

The place where the positron was emitted is close to or on the LOP, the distance to the LOP depends on the energy of the positron. The kinetic energy of the positron as well as the spatial resolution each of the detectors limit the spatial resolution of PET imaging in humans to 1–2 mm.

Integrated data of a pair of detectors in a PET camera having two to several thousands of detectors result into tomographic images. These images are reconstructed using algorithms similar to those used in X-ray CT and single photon emission computed tomography (SPECT).

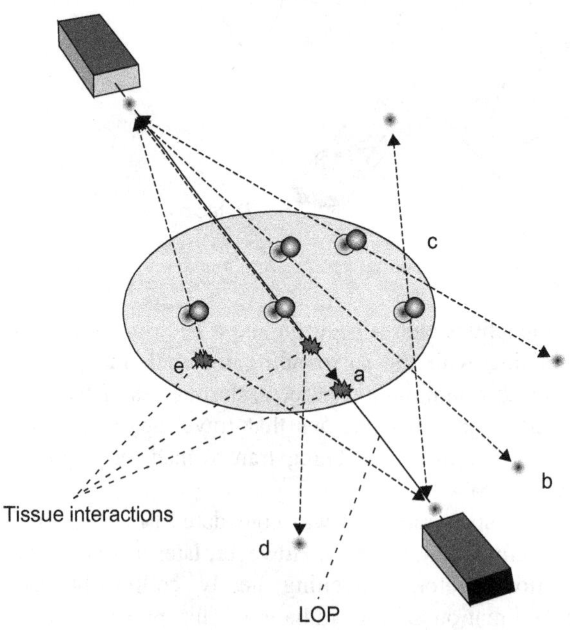

Figure 4.6 Detection disturbances aroused. (a) One of the two annihilation photons absorbed by surrounding tissue. (b) May travel in the wrong direction. (c) Two (single) photons of different annihilations may be detected simultaneously. Annihilation photons may be scattered resulting in no (d) or wrong detection (e) [4–5].

In contrast with conventional gamma camera imaging, coincidence detection excludes the necessity of collimation with a lead collimator, which significantly increases the sensitivity of PET compared to SPECT.

A pair of detectors detects not only photons that are in coincidence. Most annihilation photons do not have a coincidence detection of the second photon (Figure 4.6). It should be noted that many annihilation photons will not reach to the detector. The tissue surrounding the place of annihilation will absorb them, or they will not travel in the direction of a detector. The coincident count rate is only a fraction of the total count rate. It is also possible that two photons of different annihilations are detected (almost) simultaneously. The contribution of these random coincidences will be more significant at higher count rates. Further, photons may be scattered resulting in no or wrong LOPs. Reasonably, scattered photons and random coincidences will decrease image contrast and will hamper reliable quantitative analysis of the image data.

4.5 PET Principle

With ordinary X-ray examinations, an image is made by passing X-rays through the body from an outside source. In contrast, nuclear medicine procedures use a radioactive material called a radiopharmaceutical or radiotracer, which is injected into your bloodstream, swallowed, or inhaled as a gas. This radioactive material accumulates in the organ or area of your body being examined, where it gives off a small amount of energy in the form of gamma rays. A gamma camera, PET scanner (or probe) detects this energy and with the help of a computer creates pictures offering details on both the structure and the function of organs and tissues in your body.

Unlike other imaging techniques, nuclear medicine imaging examinations focus on depicting physiologic processes within the body, such as rates of metabolism or levels of various other chemical activities, instead of showing anatomy and structure.

Generally, in PET pictures, areas of greater intensity, called "hot spots," indicate where large amounts of the radiotracer have accumulated and where there is a high level of chemical or metabolic activity. Less intense areas, or "cold spots," indicate a smaller concentration of radiotracer and less chemical activity.

4.6 Physical Basis of PET

PET is based on the annihilation coincidence detection (ACD) of the two collinear 511 keV γ-rays resulting from the mutual annihilation of a positron and a negatron, its antiparticle (Figure 4.7). Positron–negatron annihilation occurs at the end of the positron range, when the positron has dissipated all of its kinetic energy and both the positron and the negatron are essentially at rest. The *total* positron and negatron

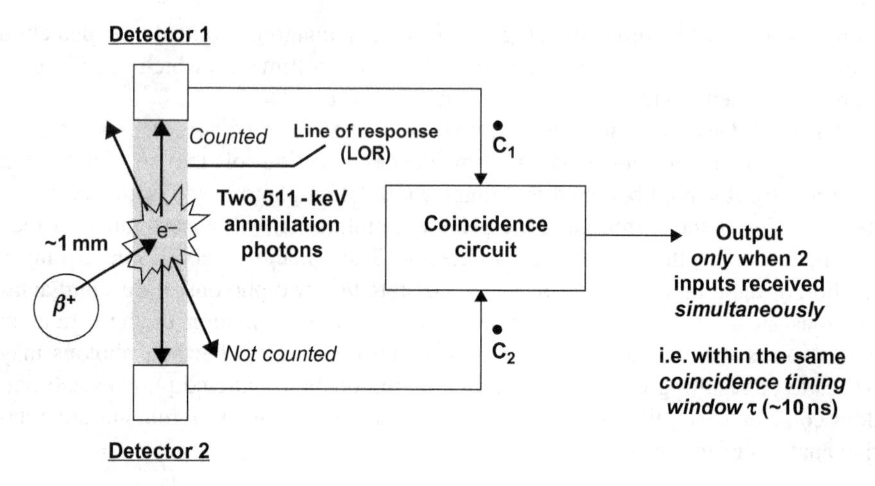

Figure 4.7 Schematic illustration of ACD. An event is counted only when each of the two 511 keV annihilation γ-rays are detected simultaneously, that is, within a time interval corresponding to the coincidence timing window τ, by the two detectors. The shaded LOR corresponds to the volume between and defined by the cross-sectional area of the coincidence detectors 1 and 2 and C_1 and C_2 are the singles count rates recorded by detectors 1 and 2, respectively.

energy is therefore 1.22 MeV, the sum of their equal rest mass energies ($E_o = 511$ keV = 0.511 MeV), and their total momentum is 0.

Accordingly, to conserve energy and momentum, the total energy of the two annihilation γ-rays must equal 1.22 MeV and their total momentum is 0.

Two equal-energy (511 keV) annihilation γ-rays traveling in opposite directions, corresponding to equal magnitude, opposite sign (positive and negative) momenta are therefore emitted.

In the parlance of ACD, each of the two annihilation photons is referred to as a "single" and the total count rate (counts per second (cps)) for the individual annihilation photons is called the "singles count rate" (Figure 4.7).

Only when signals from the two detectors simultaneously trigger the coincidence circuit, we can say that a "true coincidence event" takes place. The volume between the opposed coincidence detectors (the shaded area in Figure 4.1) is referred to as a "line of response (LOR). Thus, LORs are defined electronically.

An important advantage of ACD is that absorptive collimation is not required. As a result, the sensitivity which is measured count rate per unit activity of PET is much higher than that of Anger camera imaging by two–three orders of magnitude. We can exclude the possibility that every annihilation effect yields a counted event, but such a possibility exists. As a result, the singles count rate in PET is typically much higher than the trues count rate.

The 511 keV in energy and the simultaneity-of-detection requirements for counting of a true coincidence event are not absolute. Scintillation detectors typically have a rather coarse energy resolution—up to $\sim$30%. Usually, this parameter is determined as the percent full-width half-maximum of the 511 keV photo peak. Therefore, photons within a relatively broad energy range (e.g., 250–650 keV) can be counted as valid annihilation γ-rays. Compton-scattered annihilation γ-rays, as well as scattered and unscattered nonannihilation photons may therefore be included—they produce spurious or mispositioned coincidence events.

Each single detected photon is roughly at 1 ns $= 1 \times 10^{-9}$ s. A true coincidence event is defined as a pair of annihilation photons counted by the coincidence detectors within a time interval called the "coincidence timing window τ" (typically 6–12 ns). Such a finite timing window is necessitated by the considerations that follow. First, depending on the exact position of the positron–negatron annihilation, the annihilation photons reach the detectors at different times. This effect is very small—photons travel at the speed of light. Second, although the transit and processing of the signal pulses through the detector circuitry is rapid, it is not instantaneous. Finally, the light signal emitted by the scintillation-type detectors used in PET is emitted not instantaneously—in finite time interval, called the "scintillation decay time," of the order of 10–100 ns.

Figures 4.8 and 4.9 illustrate the various events associated with ACD of positron-emitting radionuclides.

4.7 Detectors Used in PET

Only four detector materials, all inorganic scintillators, have been widely used in PET scanners. These are the following:

1. thallium-doped sodium iodide (NaI(Tl)).
2. bismuth germanate (BGO)
3. cerium-doped lutetium
4. oxyorthosilicate (LSO(Ce))

The most important practical features of scintillation are: high mass density, effective atomic number, high light output, and speed.

A high mass number and high effective atomic number maximize the linear attenuation coefficient, and therefore the detection of radiations. Additionally, a higher atomic number crystal will have a higher proportion of photoelectric gain than Compton interactions. High light output reduces statistical uncertainty (noise or data spreading) in the scintillation and associated electronic signal. Thus, energy resolution and scatter rejection are improved. A crystal with a short scintillation decay time allows the use of a narrow coincidence timing window.

Minimal self-absorption, matching of the index of refraction of the crystal to that of the photodetector, matching of the scintillation wavelength to the light response of the photodetector and minimal hygroscopic behavior—these should be noted among other important factors.

NaI(Tl) crystals were used in the original PET scanners. Higher density and effective atomic materials, such as BGO, LSO, and GSO, have emerged as the

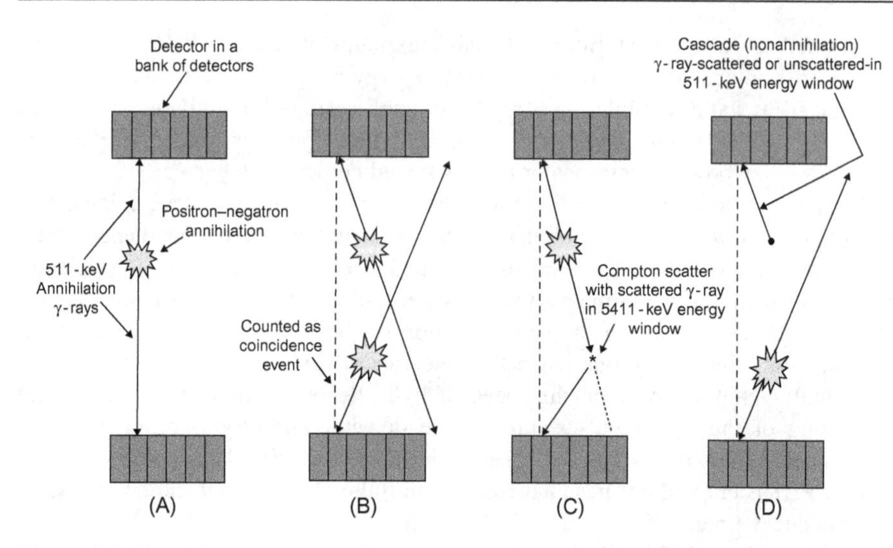

Figure 4.8 The various events associated with ACD of positron-emitting radionuclides, illustrated for two opposed banks of coincidence detectors and assuming only one opposed pair of detectors are in coincidence. (A) A true coincidence ("true") is counted only when each of the two 511 keV annihilation γ-rays for a single positron–negatron annihilation are not scattered and are detected within the timing window τ of the two coincidence detectors. (B) A random or accidental coincidence ("random") is an inappropriately detected and positioned coincidence (the dashed line) that arises from two separate annihilations, with one γ-ray from each of the two annihilations detected within the timing window τ of the coincidence–detector pair. (C) A scattered coincidence ("scatter") is a mispositioned coincidence (the dashed line) resulting from a single annihilation, with one of the γ-rays. Undergoing a small-angle Compton scatter but retaining sufficient energy to fall within the 511 keV energy window. (D) A spurious coincidence is an inappropriately detected and positioned coincidence (the dashed line) which arises from annihilation and a cascade γ-ray, scattered or unscattered but having sufficient energy to fall within the 511 keV energy window. Spurious coincidences occur only for radionuclides which emit both positron and prompt cascade γ-ray(s).

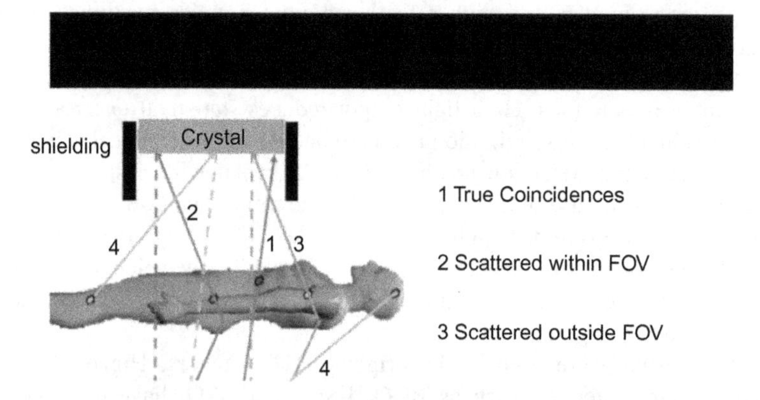

Figure 4.9 PET event types in other presentation.

detectors of choice for PET because of their greater stopping power for 511 keV annihilation γ-rays.

Materials such as GSO and LSO have a faster light output—nearly 10-fold faster than BGO.

A disadvantage of LSO is the presence of a naturally occurring long-lived radio-isotope of lutetium and an increased background count rate. Various types of PET detectors are shown in Figure 4.10.

PET procedure—how it is performed?

Usually, nuclear medicine imaging is performed on an outpatient basis but is often performed on hospitalized patients as well.

You will be positioned on an examination table. If necessary, a nurse or technologist will insert an intravenous line into a vein in your hand or arm. Depending on the type of nuclear medicine examination you are undergoing, the dose of radio-tracer is then injected intravenously, swallowed, or inhaled as a gas.

Typically, it will take approximately 60 min for the radiotracer to travel through your body and to be absorbed by the organ or tissue being studied. You will be asked to rest quietly, avoiding movement and talking.

You may be asked to drink some contrast material that will localize in the intestines and help the radiologist interpreting the study.

You will then be moved into the PET/CT scanner and the imaging will begin. You will need to remain still during imaging. The CT examination will be done first, followed by the PET scan. On occasion, a second CT scan with intravenous contrast will follow the PET scan. The actual CT scanning takes less than 2 min. The PET scan takes 20–30 min.

Total scanning time is approximately 30 min.

Depending on which organ or tissue is being examined, additional tests involving other tracers or drugs may be used, which could lengthen the procedure time to 3 h. For example, if you are being examined for heart disease, you may undergo a PET scan both before and after exercising or before and after receiving intravenous medication that increases blood flow to the heart.

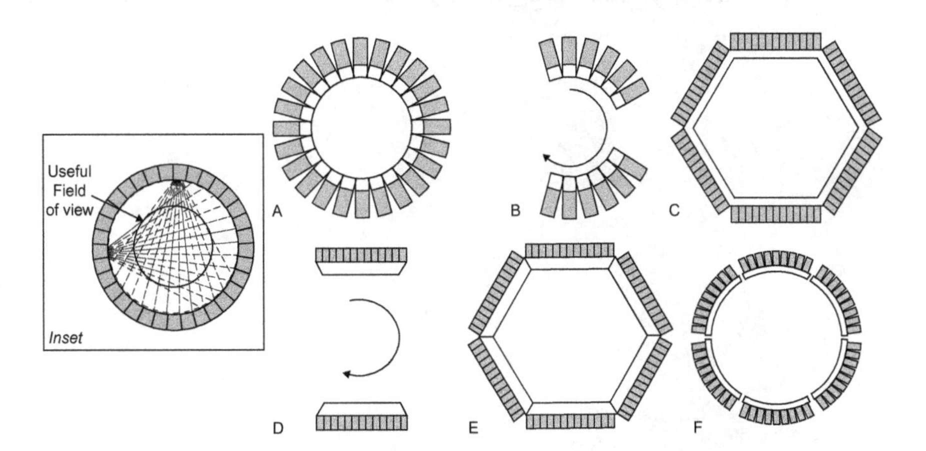

Figure 4.10 PET scanner detector configurations.

When the examination is completed, patient may be asked to wait until the technologist checks the images in case additional images are needed. Occasionally, more images are obtained for clarification or better visualization of certain areas or structures. The need for additional images does not necessarily mean there was a problem with the examination or that something abnormal was found, and should not be a cause of concern for the patient.

Nuclear medicine procedures can be time consuming and can take hours to days. During this time the radiotracer accumulate in the part of the body under study. Imaging may take up to several hours to perform. The resolution of structures of the body with nuclear medicine may not be as high as with other imaging techniques, such as CT or MRI. However, nuclear medicine scans are more sensitive than other techniques for a variety of indications, and the functional information gained from nuclear medicine examinations is often unobtainable by other imaging techniques.

Because the radioactive substance decays quickly and is effective for only a short period of time, it is important for the patient to be on time for the appointment and to receive the radioactive material at the scheduled time. Thus, late arrival for an appointment may require rescheduling the procedure for another day.

References

[1] G.L. Brownell, A history of positron imaging. Presentation prepared in celebration of 50th year of services by the author to the Massachusetts General Hospital on October 15th 1999.

[2] E. Wachaltz, History and development of PET. <CEwebsource.com>, August 15, 2013.

[3] D.L. Bailey, D.W. Townsend, P.E. Valk, M.N. Maisey (Eds.). Positron Emission Tomography, © Springer-Verlag London Limited, 2005.

[4] G. Mucllehner, J. Karp, Positron emission tomography, Phys. Med. Biol. 51 (2006) 1176.

[5] Introduction to, PET<www.Radiologyinfo.org/sitemap/modal-allas.(cxfm?moda = CT)>.

5 X-Ray Detectors

5.1 Introduction

In general, X-ray systems consist of three basic components:

1. X-ray tube and generator.
2. X-ray detector and unit that allows processing.
3. The apparatus which takes care of the geometrical arrangement of tube, patient and detector.

In order to get information about the irradiated volume, almost all imaging systems use a polychromatic X-spectrum and measure attenuation of X-rays caused by their absorption in the body. Requirements on imaging and properties of X-ray detectors are given in Table 5.1.

5.2 Classical X-Ray Generation and Detection

There is no doubt that radiography is a very popular tool in medical imaging.

A planar digital X-ray imaging device which is used in many medical applications consists of two components: (i) a radiation source that emits X-ray photons; (ii) a photodetector that generates an image from incident X-ray photons. In the case of indirect detectors, a scintillator first converts X-ray photons into light that is then detected by a photodetector. If an object is placed between source and detector, a part of the X-ray photons is absorbed. The attenuation of the incident X-ray energy varies for different types of material and is stronger for bones and weaker for soft tissue, such that a contrast can be observed in the projection image generated in the detector.

X-ray detectors started with film, screen, and film/screen systems.
X-ray radiography became digital with storage phosphor systems ("Computed Radiography").
X-ray fluoroscopy is performed with X-ray image intensifier-TV systems.
State-of-the-art X-ray imaging is done with flat-panel detectors.

X-Ray source: The design of the X-ray source is common in most X-ray devices. It consists of an evacuated vessel that contains a negatively charged cathode and a positively charged anode. A potential difference is applied in between the two components. The cathode, consisting of a spirally coiled wire, is heated up to a temperature above 2000°C (with a help of an electric current). At this relatively high

Medical Imaging Technology. DOI: http://dx.doi.org/10.1016/B978-0-12-417021-6.00005-8

Table 5.1 Requirements for Medical Image Systems

Requirements on Imaging	Requirements on Detector	Measurements Characteristics
Low noise imaging	High absorption (*Z*)	DQE
Dose reduction	High luminosity	Signal-to-noise ratio
High dynamic range	Scintillation crystal layer	
Detail contrast	High spatial resolution $< 100\ \mu m$	MTF
Dynamic imaging	High time resolution	Afterglow
Fluoroscopy	Small decay time constant	Decay time
Subsecond CT Scan	Ultralow afterglow	Charge collection time

Source: After [15]

temperature, there are electrons which move away from their nuclei and are dragged toward the anode due to the action of potential difference.

During their travel to the anode, the electrons are further accelerated such that they obtain a high kinetic energy. Obviously, the electrons hit the anode. At the same time, a small portion of their kinetic energy is converted into X-rays. Two processes are responsible for this. The first one is referred to as *bremsstrahlung radiation* (which is due to interaction of the accelerated electron with nuclei) and *characteristic radiation* (which is due to interaction with electrons in the anode).

Generally, the X-rays generated in the anode are generated in all directions, but all X-rays except for those emitted in the direction of the X-ray tube output window are absorbed (by the X-ray tube's shielding). The output window is positioned so that the X-rays that travel through it are emitted toward a detector. If an object is inserted into the path between emitter and detector, the X-ray photon can either be scattered, absorbed, or not interacted with the object at all. The contrast in the image that is generated in the detector is due to the difference in the amount of X-ray photons that are absorbed or scattered by the different tissues in the body and hence do not reach the detector.

5.3 Key Parameters

First of all let us consider what are the important parameters for medical X-ray detector.

5.3.1 Image Receptor Format

It is well known that X-rays always propagate along straight lines even if they are scattered. It should be stressed on that radiation with a broad spectrum used for medical purposes cannot be deflected or focused. Therefore, in a medical application the radiation receptor has to be of at least the same size as the human body or the organs under investigation. This fact calls for a detector based on large area electronics containing extended semiconductor layers.

Hydrogenated amorphous silicon (a-Si:H) has proven to be the best material available for this purpose. Other thin film materials such as amorphous selenium (a-Se) or cesium iodide (CsI) are also used for X-ray absorption.

If large formats are required, in conventional radiology film–screen systems or storage phosphor plates up to $35 \times 43\ cm^2$ are used. XRII-TV systems cover the range from 17 to 40 cm diameter input screen.

The charge-coupled devices (CCDs) can be used for rather small areas. A CCD sensor is coupled to a fiber-optic plate (FO) which serves as a radiation shield. The FO is covered with a scintillator, for instance CsI. Such a detector with 40 μm pixel size is used for dental imaging. However, CCD technology is limited in size. The largest CCD of the world produced in quantity is $49 \times 86\ mm^2$ in size (i.e., 4 K × 7 K pixels of 12 μm). It is used in a sensor for mammography biopsy.

Because most of the applications mentioned above require larger detectors than CCDs, a-Si-based detectors have been developed.

5.3.2 Spatial Resolution

Obviously the required spatial resolution depends on the kind of objects a radiologist wants to investigate.

Soft tissue can be imaged with 1–2 lp/mm, fine bone structures require >3 lp/mm to be precisely imaged.

Resolutions >5 lp/mm are demanded for mammography detectors–there microcalcifications of some 100 μm shall be imaged.

The same is true for dental applications.

The highest spatial resolution of an X-ray detector is determined by different physical processes but there is no need to go into details. Reasonably those real detectors are not idealized planes but have a nonzero thickness. Therefore, X-rays impinging the detector under an angle different from the normal direction are smeared laterally.

In a case when a scintillator is used as an absorber, light propagation in the scintillator as well as the optical coupling between scintillator and photodiodes should be considered. For radiographic applications, CsI with a thickness of 400–600 μm has proven to be a good compromise. For a very high resolution, required, 100–200 μm thickness is more suitable.

A directly converting semiconductor has an extremely high spatial resolution—and this is their advantage. To author's best knowledge, there is no optical effects which can disturb the image. Lateral diffusion of charge carriers may be the only one that can reduce resolution.

If we analyze the data for a-Se layer of 500 μm, a drift length far exceeds the diffusion length at normal applied. That is why the a-Se layer is suitable for spatial resolutions even >10 lp/mm.

Reduction of spatial resolution arose in the case of the array structure of the solid-state detectors. The resulting signal is averaged over the active part of the pixels, i.e., the area of the photodiodes or the electrodes of the directly converting semiconductor layer. Moreover, sampling is performed at discrete points which may lead to aliasing effects. High spatial frequency objects are imaged with so-called pseudo-sharpness.

5.3.3 Frame Rate

As a rule, solid-state detectors operate in real time. Real time in cardiac imaging means 30 Hz (or even 60 Hz in pediatry). Fluoroscopy is performed at 15–30 Hz; if needed (in order to reduce X-ray dose), it can be performed at ~10 Digital subtraction angiography (DSA) is done at 2–8 Hz. In radiography, mammography, or dental imaging, even one image every 30 s can still be called real time, compared to the time required for film exposure, transport, and development.

Two main intrinsic factors limit the performance of an imager panel in the time domain: image lag and incomplete charge transfer. The latter is due to the on-resistance of the readout switches which leads to some 10 μs required to assure that at least 99% of the signal can be transferred to the readout amplifiers.

Image lag has its origin in both, scintillator and a-Si readout matrix. The scintillator afterglow can be reduced to an acceptable level by selecting the appropriate material, controlling deposition parameters, doping level, and postdeposition treatment (annealing).

Trap filling and emission of charge carriers from traps cause lag from a-Si. At the same time, the concentration (density) of traps responsible for this drawback can be reduced by minimizing the geometrical device volume. The a-Si bulk density of states has already reached a concentration of $\approx 10^{15}$ cm^{-3} in state-of-the-art material. Lag effects can be reduced notably by keeping the traps always filled. This filling can be accomplished by a flashed backside illumination of the detector.

5.4 The Direct and Indirect Mode of Operation

X-ray detectors may be considered as a relay of components that serve to stop the X-rays in a material and produce an initial signal that may then be amplified or converted to another form of energy, e.g., in an electrical signal. The energy converter may be a light-emitting phosphor layer, whose signal is amplified by an image intensifier before being captured on a CCD. That process is an example of indirect detection. That is, signal quanta produced in the X-ray stopping medium (visible light in the phosphor) differ from the quanta ultimately recorded (charge in the CCD).

In a direct detector, the energy converter and the final imager are the same component—a strip of X-ray film.

Phosphors are still the basis of most X-ray imagers, which typically operate indirectly. The desired phosphor characteristics logically follow from the imaging requirements. At the same time, it should be noted here that only a small number of the commonly used phosphors, e.g., thallium-doped CsI and terbium-doped gadolinium-oxide sulfide, meet most of detector requirements. Among these high quantum efficiency and a high spatial resolution should be mentioned first of all.

5.5 Principles of the Direct-Conversion Digital X-Ray Image Detector

A systematic historical overview of the evolution of digital radiography (DR) is shown in Table 5.2. Experimental digital subtraction angiography was first described in 1977 and introduced into clinical use as the first digital imaging system in 1980. For general radiography, X-ray images were first recorded digitally with cassette-based storage-phosphor image plates, which were also introduced in 1980.

The first DR system, which appeared in 1990, was the CCD slot-scan system. In 1994, investigations of the selenium drum DR system were published. The first flat-panel detector DR systems based on amorphous silicon and amorphous selenium were introduced in 1995. Gadolinium-oxide sulfide scintillators were introduced in 1997 and have been used for portable flat-panel detectors since 2001. The latest development in DR is dynamic flat-panel detectors for digital fluoroscopy and angiography.

The most obvious advantage of digital detectors is that they allow implementation of a fully digital picture archiving and communication system, with images stored digitally and available anytime. Thus, distribution of images in hospitals can be achieved electronically by means of web-based technology without the risk of losing images. Other advantages include higher patient throughput, increased dose efficiency, and the greater dynamic range of digital detectors with possible reduction of X-ray exposure to the patient.

In this chapter, we provide an overview of the DR systems currently available for general radiography. In so doing, we describe the physical principles of DR and discuss and illustrate different systems in terms of detectors, image processing, image quality criteria, and radiation exposure issues. Future technologies and perspectives in DR are also discussed.

There has been much work over the last decades on the development of solid-state flat-panel detectors, also known as active matrix flat-panel imagers (AMFPI). Unlike the XRII, these systems are thin, produce negligible spatial distortions and

Table 5.2 Timetable of Developments in DR

Year	Development
1977	Digital subtraction angiography
1980	Computed radiography (CR), storage phosphors
1987	Amorphous selenium-based image plates
1990	Charge-coupled device (CCD) slot-scan direct radiography
1994	Selenium drum DR
1995	Amorphous silicon–cesium iodide (scintillator) flat-panel detector
1995	Selenium-based flat-panel detector
1997	Gadolinium-based (scintillator) flat-panel detector
2001	Gadolinium-based (scintillator) portable flat-panel detector
2001	Dynamic flat-panel detector fluoroscopy–digital subtraction angiography

are insensitive to magnetic fields. In principle, panels with sufficiently small pixels, or *detector elements*, also have the potential for substantially improved spatial resolution in comparison with the XRII because there are considerably less conversion stages involved in the image detection process. At present, flat-panel X-ray imagers are now widely used in digital X-ray imaging with applications in medical and non-medical imaging.

There is still a need for an ideal digital X-ray imaging system. What would such a system provide? First, it should provide a reduction in X-ray exposure or dose. Second, the image should be of a high quality and be available almost immediately such that it would be available for use in real-time imaging (e.g., fluoroscopy). It would be so designed as to minimize cost and thus conveniently incorporated into medical systems. The "ideal" system would record the X-ray image directly onto a computer where it could be read, stored, and analyzed. Such a system can be achieved using a flat-panel X-ray image detector (sensor).

Research over the past decades has indicated that the most practical flat-panel system would be based on a large area integrated circuit. Such large integrated circuits or active matrix arrays (AMAs) have been developed as the basis for large area displays. AMAs based on hydrogenated amorphous silicon (a-Si:H) thin film transistors (TFTs) have been shown to be practical image addressing systems. They may be converted into X-ray sensitive imaging devices by adding a thick (~1 mm) X-ray detecting medium. Systems using either a phosphor or a photoconductor are possible. The AMA used for image addressing and readout in a flat-panel X-ray image detector consists of many single pixels, each of which represents a corresponding pixel of image. Each pixel has some charge that is proportional to the X-ray radiation that it receives. To generate this signal charge, either a phosphor is used to convert the X-rays to visible light which in turn is detected with a p-i-n photodiode at the pixel (indirect) or an X-ray photoconductor converts the incident X-rays to charge (direct) in one step. Several manufacturers and academic researchers have used the indirect approach. The physical form of the system would be similar to a film/screen cassette so that it would easily fit into current medical X-ray systems as illustrated in Figure 5.1.

At present, essentially two methods have been adopted for DR (Table 5.3; Figure 5.2). Both are based on the use of phosphors. This means that both involves indirect conversion from X-ray photon to a detectable charge signal. The first is the digitalization of a signal from a video camera optically coupled to an X-ray image intensifier (CsI phosphor). The second is the photostimulable phosphor system, commonly called computed radiography (CR) system, that captures the latent image within a storage phosphor layer, which is subsequntly readout with a laser scanner. The intensifier system permits instantaneous readout. It is very bulky while the CR system, like film/screen, requires carrying the cassette from a loading/unloading station to the patient examination room and back. Neither of these indirect-conversion phosphor-based system has adequate image quality for all applications. The need for a DR system that reads out images electronically and directly and with better image quality remains.

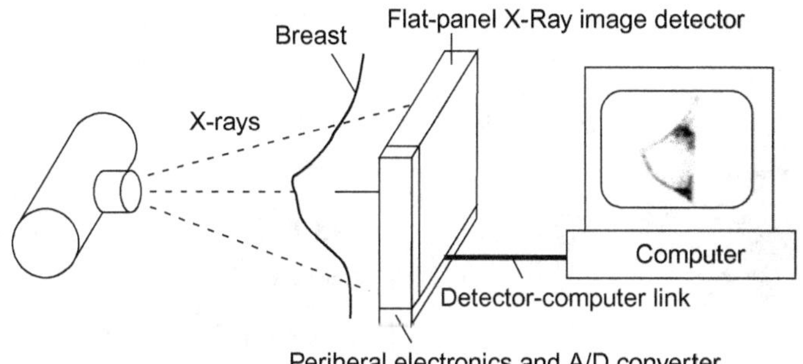

Figure 5.1 Schematic illustration of a flat-panel X-ray image detector for digital mammography. Connection from the detector to a local or distant computer is a convenient communications link [1,2].

Table 5.3 Conversion of X-Rays to Images—Comparison of Different Methods

Regime	Detection method	Detector	Conversion from X-Rays to Image
Digital	Direct	Flat-panel detector (amorphous aelenium)	X-rays→image
Digital	Indirect	Flat-panel detector (fluorescent material + + photodiode)	X-rays→light→image
Digital	Indirect	I.I.* + TV camera	X-rays→light→image
Digital	Indirect	Imaging plate	X-rays→latent image→light→image
Analog	Indirect	I.I.* + cine film	X-rays→latent image→light→image
Analog	Indirect	Intensifying screen + +X-ray film	X-rays→latent image→light→image

I.I.*—Image Intensifier

The majority of commercial AMFPIs are indirect-conversion detectors, in which X-ray photons strike a scintillator such as CsI and generate optical photons which then interact with a photosensor (usually an amorphous silicon photodiode), in turn producing electron–hole-pairs (EHP) that are capacitively stored prior to being electronically processed. The process of detecting the photon-generated charge from each detector element (referred to as readout) produces a digital image which represents the original distribution of X- rays incident at the imager's surface. In another class of detectors known as direct-conversion detectors, X-rays interact with a photoconductor, usually amorphous selenium (a-Se), and directly generate EHPs, which follow the parallel field lines in the presence of an electric field prior to being read out. Because there is no intermediary optical stage to contribute to

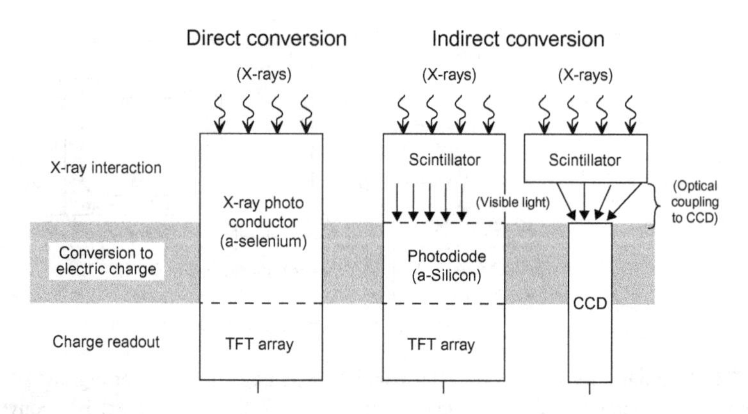

Figure 5.2 Electronically readable detectors.

blurring, these systems have the important advantage of providing superior spatial resolution compared to indirect AMFPIs, however, the photoconductor needs to be thick enough (~1000 μm) to yield a reasonable quantum efficiency at radiographic X-ray energies (20−150 kV). The very high spatial resolution of direct-conversion a-Se detectors has recently increased their use in digital mammographic imaging systems. Direct-conversion detectors are also being considered for use in tomosynthesis, in which a series of breast radiographs are acquired from different angles and reconstructed into a series of slices. In contrast with mammography, the image slices produced by tomosynthesis are largely immune to structural noise, which is the noise introduced into an image due to X-ray attenuation in overlapping anatomical structures. Most important applications of flat-panel X-ray imagers (FPXIs) are in medical imaging such as mammography, chest radiology, angiography, and fluoroscopy.

Conversion of X-rays to images, parameters for digital imaging and properties typical for selected X-ray photoconductors used in large area applications are summarized in Table 5.3. Figures 5.3−5.5 exemplifies FPXI, a simple pixel with TFT and typical X-ray images, correspondingly.

Recent research carried out concurrently at Sunnybrook Health Science Centre [1−4], DuPont [5,6], and Philips [7] has shown clearly that one of the most promising flat-panel digital radiographic systems is based on using a large area thin film transistor (TFT) AMA with an electroded X-ray photoconductor. An important first step in the flat-panel thin film technology was the development of flat-panel TFT displays which matured as the fabrication and doping of large area hydrogenated amorphous silicon (a-Si:H) films became technologically possible in the early 1990s [8]. Once large area flat panels with small pixel sizes became available at the component level it was only a matter of time before an X-ray photoconductor such as amorphous selenium would be used to directly convert X-ray images to a charge distribution stored on the pixels of a flat panel. Amorphous selenium photoconductor-based flat-panel X-ray detectors are now at a stage that can provide

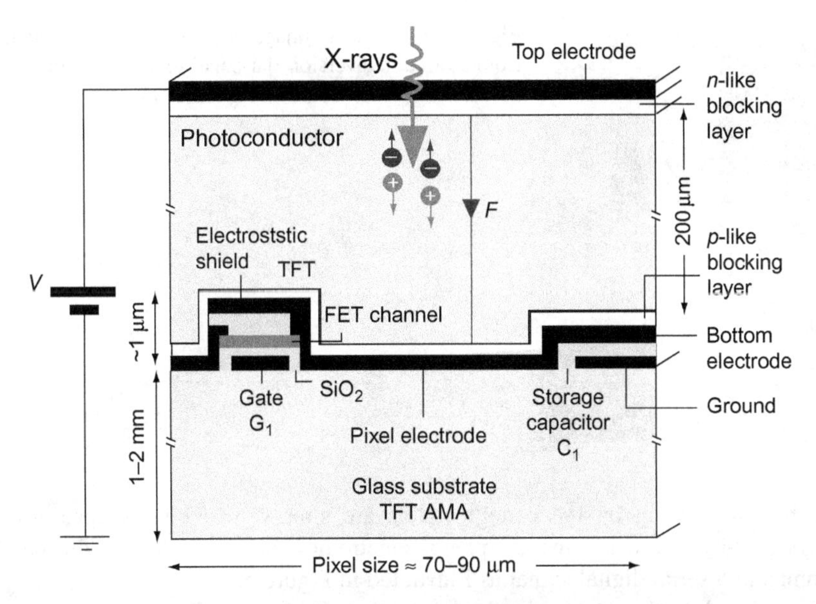

Figure 5.3 Cross section of a single pixel (i,j) with a TFT showing the accumulation of X-ray generated charge on the pixel electrode and hence the storage capacitance C_{ij}. The top electrode (A) on the photoconductor is a vacuum-coated metal (e.g., Al). The bottom electrode (B) is the pixel electrode that is one of the plates of the storage capacitance (C_{ij}) [3].

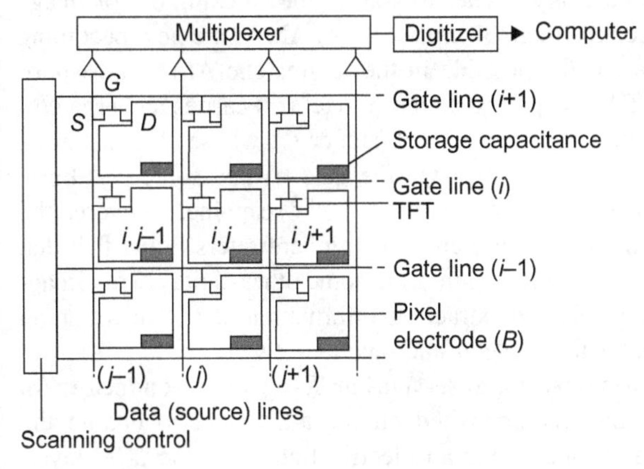

Figure 5.4 TFT AMA for use in X-ray image detector with self-scanned electronic read out. The charge distribution residing on the panel's pixels are simply read out by scanning the arrays row-by-row using the peripheral electronics and multiplexing the parallel columns to a serial digital signal [3].

excellent images and are being employed in various X-ray imaging application—medical and nonmedical. The combination of an AMA and an X-ray photoconductor then constitutes a direct-conversion X-ray image detector; a term coined by DuPont (the terms X-ray sensor and detector have been used interchangably in the literature). Direct conversion here refers to the fact that the X-ray photons are directly converted to charges and detected vis-à-vis an intermediate conversion, via a phosphor, to photons (light) and then from photons to charges. The charge

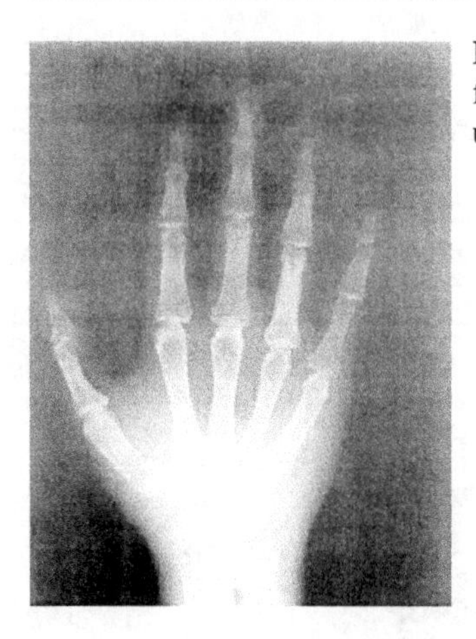

Figure 5.5 An X-ray image of a phantom hand obtained from a direct-conversion flat-panel image detector that uses a stabilized a-Se photoconductor [3].

distribution residing on the panel's pixels are simply read out by scanning the arrays row-by-row using the peripheral electronics and multiplexing the parallel columns to a serial digital signal as illustrated in Figure 5.4.

The signal is then transmitted to a computer system. The system is simple, inherently digital and has so many advantages that it has now become a major contending choice in DR [3].

An AMA consists of millions of individual pixel electrodes connected by TFTs (one for each pixel) to electrodes passing over the whole array to subsidiary electronics on the periphery. The TFTs act as switches to control the clocking out of image charge a line at time. Very large area (e.g., 30 cm × 30 cm) AMAs are now becoming available and even larger ones will be possible in the future. The AMA consists of M × N (e.g., 1280 × 1536) TFT-based pixels and each pixel (i,j) carries a charge collection electrode B connected to a signal storage capacitor C_{ij} whose charge can be read out by properly addressing the TFT (i,j) via the gate (i) and source (j) lines. External readout electronics and software, by proper self-scanning, converts the charges read on each C_{ij} to a digital image. Self-scanning here refers to the fact that no external means, such as a scanning laser beam as in some other digital X-ray imaging systems, is used to scan the pixels and extract the information [2,3]. The scanning operation is part of the flat-panel detector electronics and its software.

The a-Se layer is coated onto the AMA to serve as an X-ray photoconductor. An electrode (labeled A) is subsequently deposited on the a-Se layer to enable the application of a biasing potential and hence an electric field E in the a-Se layer. The EHP that are generated in the photoconductor by the absorption of an X-ray photon travel along the field lines. Electrons are collected by the positive bias electrode (A) and holes accumulate on the storage capacitor C_{ij} and thereby provide a charge signal ΔQ_{ij} that can be read during self-scanning.

Each pixel electrode carries an amount of charge ΔQ_{ij} that is proportional to the amount of incident X-ray radiation by virtue of the X-ray photoconductivity of the photoconductor over that pixel. All TFTs in a row have their gates connected whereas all TFTs in a column have their sources connected. When gate line i is

Table 5.4 Parameters for Digital X-Ray Imaging Systems

Clinical Task	Chest Radiology	Mammography	Fluoroscopy
Detector size	35 cm × 43 cm	18 cm × 24 cm	25 cm × 25 cm
Pixel size	200 μm × 200 μm	50 μm × 50 μm	250 μm × 250 μm
Number of pixels	1.750 × 2.150	3.600 × 4.800	1000 × 1000
Readout time	~1 s	~1 s	1/30 s
X-ray spectrum	120 kVp	30 kVp	70 kVp
Mean exposure	300 μR	12 mR	1 μR
Exposure range	30–300 μR	0.6–240 mR	0.1–10 μR
Radiation noise	6 μR	60 μR	0.1 μR

activated, all TFTs in that row are turned "on" and N data lines from $j = 1$ to N then read the charges on the pixel electrodes in row i. The parallel data are multiplexed into serial data, digitized, then fed into a computer for imaging. The scanning control then activates the next row, $i + 1$, and all the pixel charges in this row are then read and multiplexed, and so on until the whole matrix has been read from the first to the last row (Mth row). It is apparent that the charge distribution residing on the panel's pixels is simply read out by self-scanning the arrays row-by-row and multiplexing the parallel columns to a serial digital signal. This signal is then transmitted to a computer system. Figure 5.5 shows a typical X-ray image of a phantom hand obtained by a flat-panel X-ray image detector using a-Se photoconductor as developed by Sterling Diagnostic Imaging (formerly DuPont). The resolution is determined by the pixel size which in present experimental image detectors is typically ~150 μm but is expected to be as small as 50 μm in future detectors to achieve the resolution necessary for mammography.

Undoubtedly, it is an interesting feature of the a-Se based flat-panel X-ray sensor—this technology has been made possible by the use of two key elemental amorphous semiconductors: a-Si:H and a-Se. Although their properties are different, both can be readily prepared in large areas which is essential for an X-ray image detector. It will be impractically difficult and expensive to develop a large area detector using a single crystal technology [3].

Any flat-panel X-ray image detector design must first consider the required specifications based on the clinical need of the particular imaging modality, e.g., mammography, chest radiology, and fluoroscopy.

Table 5.4 summarizes the specifications for flat-panel detectors for chest radiology, mammography, and fluoroscopy.

5.6 The Ideal X-Ray Photoconductor

The initial photoconductor that has been tried and found to be highly successful, just as in the case of electrophotography, has been amorphous selenium. The initial choice was based on a number of unique features including a-Se being an easy material to coat a thick film over large areas and still maintain its properties uniformly. It is also highly X-ray sensitive which is the key requirement. Amorphous selenium in the form of large photoreceptor plates was originally used in xeroradiography which is essentially the photocopying of a body part using X-rays instead of light [9]. However, this system suffers from the difficulties and noise associated with the powder development technique. Xeroradiography is no longer competitive because of the toner readout method, not the underlying properties of the a-Se photoconductor [10]. By replacing the toner readout with an electrostatic readout, a-Se has again become the basis of a clinical imaging system and commercial interest in a-Se has been revieved.

The flat-panel X-ray image detectors that use an a-Se photoconductor have been demonstrated to provide excellent images. Amorphous selenium may not be the only choice just as present xerographic photoreceptors eventually moved from a-Se to new human-made (organic) photoconductors that had distinct economic advantages [11] (Table 5.5).

It is therefore instructive to identify what constitutes a nearly perfect photoconductor in search of improving their performance or finding better materials. Ideally the photoconductive layer should possesses the following material properties [3]:

a. The photoconductor should have as high an intrinsic X-ray sensitivity as possible, that is, it must be able to generate as many collectable (free) EHP as possible per unit of incident radiation.

b. Nearly all the incident X-ray radiation should be absorbed within a practical photoconductor thickness to avoid unnecessary exposure of the patient.

c. There should be no dark current. This means the contacts to the photoconductor should be noninjecting and the rate of thermal generation of carriers from various defects or states in the band-gap should be negligibly small. In the case of electrostatic photoconductors, the dark discharge of the surface potential (in the absence of X-rays) should be negligible.

d. There should be no bulk recombination of electrons and holes as they drift to the collection electrode; EHP are generated in the bulk of the photoconductor.

e. There should be no deep trapping of EHP which means that, for both electrons and holes, the schubweg $\mu\tau E \gg L$ where μ is the drift mobility, τ is the deep trapping time (lifetime), E is the electric field, and L is the layer thickness.

f. The longest carrier transit time, which depends on the smallest drift mobility, must be shorter than the access time of the pixel and interframe time in fluoroscopy. The transit time in the electrostatic readout is not constant because the surface voltage is photo-discharged by X-ray radiation [12].

g. The above should not change or deteriorate with time and as a consequence of repeated exposure to X-rays, i.e., X-ray fatigue and X-ray damage should be negligible.

h. The photoconductor should be easily coated onto the AMA panel, e.g., by conventional vacuum techniques. Special processes are generally more expensive. In the case of electrostatic readout, the photoconductor should be coatable on a large area conducting substrate.

The large area coating requirement in (*h*) over areas typically 30 cm × 30 cm, or greater, rules out the use of X-ray sensitive crystalline semiconductors which are difficult to grow in such large areas. Various polycrystalline semiconductors such

Table 5.5 Properties Typical for Selected X-Ray Photoconductors Used in Large Area Applications [15]

Photoconductor, State, Preparation	δ at 20 keV δ at 60 keV	E_geV	W ± eV	Electron $\mu_e\tau_e$ (cm^2/V)	Hole $\mu_h\tau_h$ (cm^2/V)
Stabilized a-Se Amorphous Vacuum deposition	49 μm 998 μm	2.2	45 at 10 V/μm 20 at 30 V/ μm	$3\times10^{-7}-10^{-5}$	$10^{-6}-6\times10^{-5}$
HgI_2 Polycrystalline PVD	32 μm 252 μm	2.1	5	$10^{-5}-10^{-3}$	$10^{-6}-10^{-5}$
HgI_2 Polycrystalline SP	32 μm 252 μm	2.1	5	$10^{-6}-10^{-5}$	10^{-7}
$Cd_{95}Zn_{0.5}Te$ Polycrystalline Vacuum deposition	80 μm 250 μm	1.7	5	2×10^{-4}	3×10^{-6}
PbI_2 Polycrystalline Normally PVD	28 μm 259 μm	2.3	5	7×10^{-8}	2×10^{-6}
PbO Polycrystalline Vacuum deposition	12 μm 218 μm	1.9	8–20	5×10^{-7}	small
TlBr Polycrystalline Vacuum deposition	18 μm 317 μm	2.7	6.5	small	$1.5-3\times10^{-6}$

as $Zn_yCd_{1-y}Te$, PbI_2, may be prepared in large areas, but their main drawback is the adverse affect of grain boundaries in limiting charge transport and, further, the high substrate and annealing temperatures required to optimize the semiconductor properties. Organic photoconductors that currently dominate the xerographic photoreceptor industry and can be cheaply prepared in large areas are useless because they do not satisfy (a) and (b). On the other hand, amorphous semiconductors such as a-Se, a-As_2Se_3, and a-Si:H are routinely prepared in large areas for such applications as xerographic photoreceptors and solar cells and are therefore well suited for flat-panel X-ray detector application. Among the three, a-Se is particularly well poised because it has a much greater X-ray absorption coefficient than a-Si:H, due to greater *Z* (atomic number), and it possesses good charge transport properties for both holes and electrons compared with a-As_2Se_3 in which electrons become trapped and hole mobility is much smaller. In addition, dark current in a-Se is much smaller than that in a-As_2Se_3. Due to its commercial use as an electrophotographic photoreceptor, a-Se is one of the most highly developed photoconductors [13]. It can be easily coated as thick films (e.g., 100–150 μm) onto suitable substrates by conventional vacuum deposition techniques and without the need the substrate temperature beyond 60–70°C. Its amorphous state maintains uniform characteristics to very fine scales over large areas. A large area detector (e.g., at least 24×18 cm^2 for mammography) is esential in radiography since the lack of a practical means to focus X-rays necessitates a shadow X-ray image which is larger than the body part to be imaged.

5.7 Intrinsic Resolution of X-Ray Photoconductors

Photoconductors that directly convert the X-ray radiation to EHPs have a number of distinct advantages, one of which is their intrinsic high resolution. The resolution of an imaging device is specified in terms of its modulation transfer function (MTF), which is the relative response of the system as function of spatial frequencies. The higher the MTF, the better the resolution can be. It is instructive to examine the intrinsic resolution of a photoconductor-based detector. Consider an electroded a-Se layer that has been biased to establish field F in the photoconductor and assume that the pixel size is negligibly small. X-ray absorbed in the photoconductor release EHPs. Holes are drawn to the top elecrtode and become neutralized; electrons accumulate on the storage capacitance, so forming the latent charge image.

The lateral spreading of information and, hence, the loss of resolution in a photoconductor-based detection system can be attributed to a number of causes. The extention of Que and Rowlands ideas [14] to the present electroded system suggests the following causes for the loss of resolution [2]:

1. The range of primary electrons generated by photoelectric effect.
2. Reabsorption of characteristics K-fluorescent X-rays away from the original photoelectric absorption site.
3. Reabsorption of Compton scattered photons.
4. Internal diffusion of drifting X-ray photogenerated charge carriers as they traverse the photoconductor thickness.
5. Internal spreading due to the internal field arising from injected carriers, i.e., space charge effects that arise as a result of the charge of the injected carriers or, in other words, Coulombic repulsion between the drifting charge of the same sign.
6. Induced charges in neighboring pixels due to trapped charges in the photoconductor.
7. Bulk space charge due to trapped carriers perturbing the field which modifies the photogeneration process and changes the charge carrier transport and collection characteristics.
8. Geometric blurring due to the oblique incidence of X-rays and finite-photoconductor thickness.

These effects are schematically illustrated in Figure 5.6. Que and Rowlands found that the range of primary electrons generated by the absorbed X-ray photon and the oblique X-ray incidence effects limits the resolution of the a-Se photoconductor. They were able to conclude that the inherent resolution of the a-Se photoconductor system is far superior to that of the CsI-based columnar phosphor system. The range of primary electron that is generated by an absorbed photon depends on its energy and the density of material. This range is typically $\sim 1-3\ \mu m$ at 10–30 keV and $\sim 10-30\ \mu m$ at 50–100 keV. K-fluorescent X-rays may be released after the interaction of an X-ray photon with the K shell of an atom. The fluorescent X-rays are released isotropically and can be reabsorbed at a point distant from their creation, thus, giving rise to a characteristic type of blurring above the K edge of selenium. Geometrical blurring arises when X-rays are oblique incident. Since photons are absorbed at different depths, they give a different response at the collection pixels depending on the depth of absorption. For an a-Se

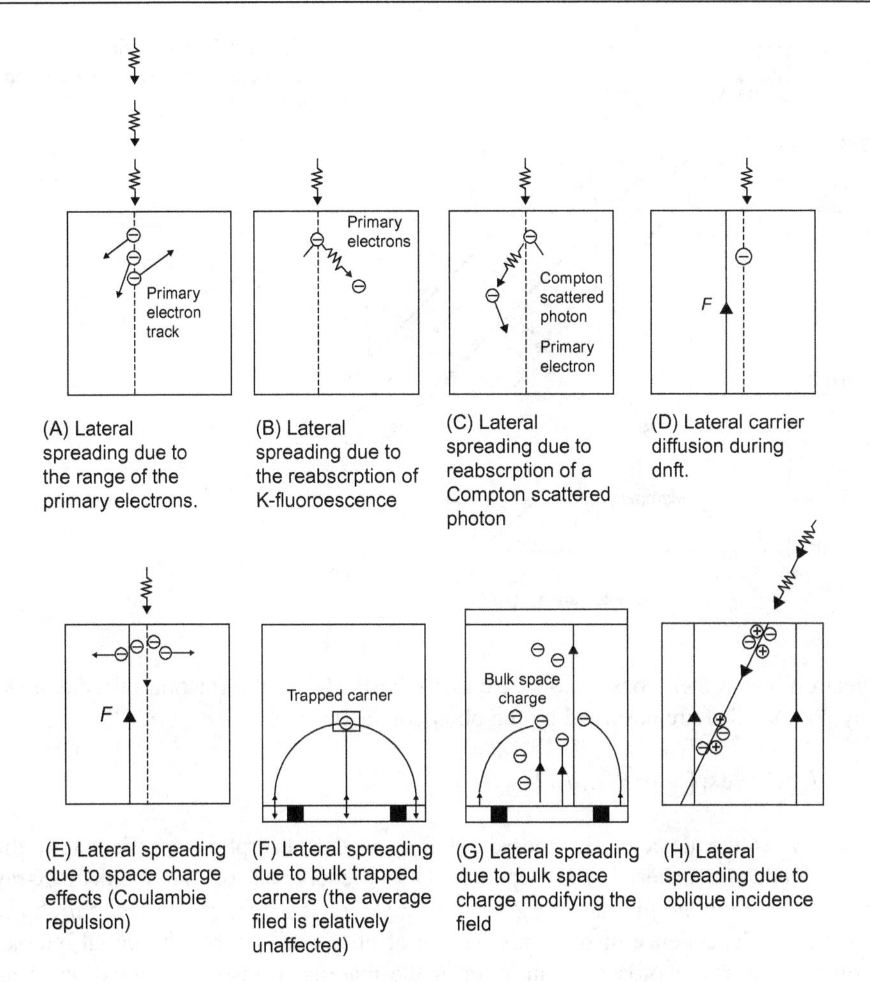

Figure 5.6 Various mechanisms that can lead to the broadening of the image and, hence, a loss of resolution in a photoconductor-based X-ray imaging system [2].

photoconductor of thickness 200–1000 μm and for the largest angle of incidence of the order of 15°, the blurring can be of the order of 50–250 μm (a significant amount compared to pixel size).

5.8 Absorption, Photoconductor Thickness, and Carrier Schubwegs

The photoelectric absorption coefficient α represents the interaction of the incident X-ray photon with the atoms in the material in producing an energetic primary

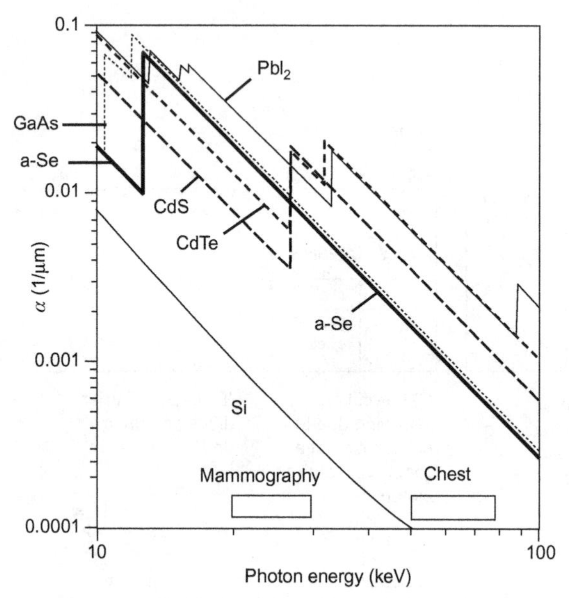

Figure 5.7 Absorption coefficient (1/μm) versus photon energy (keV) [2,3].

electron which then goes on to create many EHP. If Φ is the fraction of incident X-ray photons that are absorbed by the photoconductor then

$$\Phi = 1 - \exp[-\alpha(E, Z, d)L]$$

where L is the detector thickness, $\alpha(E,Z,d)$ is the absorption coefficient of the photoconductor material and depends on the energy E of the incident X-ray photons and the atomic number Z and density d of the material. Figure 5.7 shows the energy dependence of α for a selection of photoconductors. The initial interaction of an X-ray photon with an atom of the material leads to the emission of an energetic electron from an inner core, such as the K shell, into the conduction band. This is the photoelectric effect and the α versus photon energy curves in Figure 8.6 corresponds to the sharp vertical edges. Each shell marks an onset of absorption and as the energy increases the absorption coefficient decreases typically as E^{-n} where, for example, $n \approx 3$. The absorption coefficient increases with atomic number Z of the material, typically as $\alpha \sim Z^n$, where $n \approx 3-5$. The low Z is the primary reason for inexpensive organic semiconductors, and also a-Si:H, being excluded as X-ray photoconductor candidates (Figure 5.8).

The minimization of dosage requires α to be such that the most of the radiation is absorbed within L or $1/\alpha < L$. The required photoconductor thickness has to be several times the absorption depth $1/\alpha$ which means that it depends on the photon energy hence the particular imaging application and the location of the K and L edges. For mammography with 20 keV, the required a-Se thickness, taken as $2/\alpha$, is about 100 μm whereas it is about 2000 μm for chest radiology with a mean photon energy of 60 keV. For comparison, the corresponding thicknesses for a CdTe

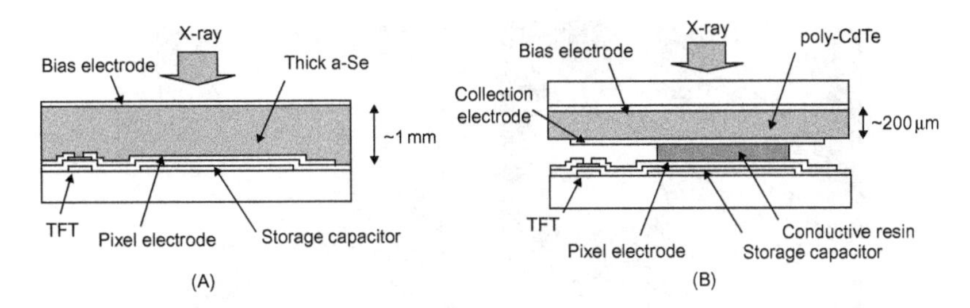

Figure 5.8 Panel structure for (A) a-Se and (B) CdTe.

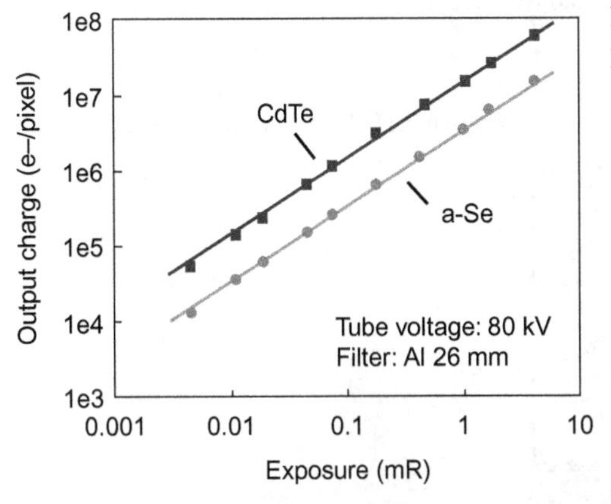

Figure 5.9 X-ray sensitivity of prototypes.

detector are 160 and 500 μm, respectively. The distinct advantages of a-Se is the fact that it can be conveniently prepared as uniform thick layers over large areas by simple vacuum deposition without harming the underlying AMA substrate. Growth of single crystals of GaAs, CdTe, CdZnTe, ThBr, and other potential X-ray photoconductors is technologically limited to small areas not suitable for large area detection; typically diameters not eceeding a few inches. Although it is potentially possible to prepare large area polycrystalline layers of PbO and PbI_2, the former has stability problems and the latter has limited sensitivity due to charge carrier schubweg limitation. Figures 5.9–5.11 compare some parameters for two solid-state Flat panel detectors (FPDs).

An important parameter that controls the charge collection efficiency η of X-ray generated charge carriers is the schubweg per unit thickness, $s = \mu\tau F/L$. The actual collection efficiency depends in a complicated way not only on the electron and hole schubwegs but also on the attenuation coefficient of the material that determines the distribution of photogenerated carriers. Nonetheless, $s \gg 1$ for both electrons and holes is still a necessary condition for near-perfect collection: $\eta \approx 1$. In addition, the relative importance of whether hole or electron schubweg is the primary controlling factor in the collection efficiency depends on the polarity applied to the radiation receiving (front) electrode and the absorption depth per unit

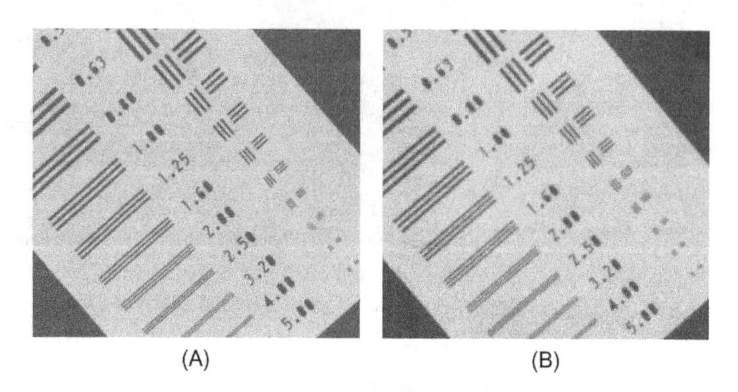

Figure 5.10 Resolution pattern: (A) a-Se and (B) CdTe.

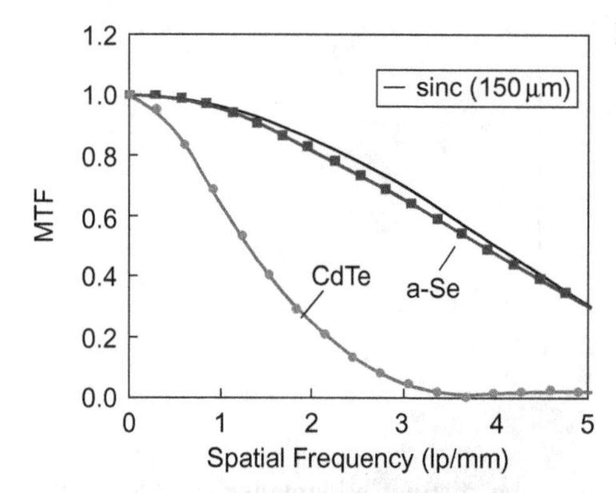

Figure 5.11 MTF for prototypes.

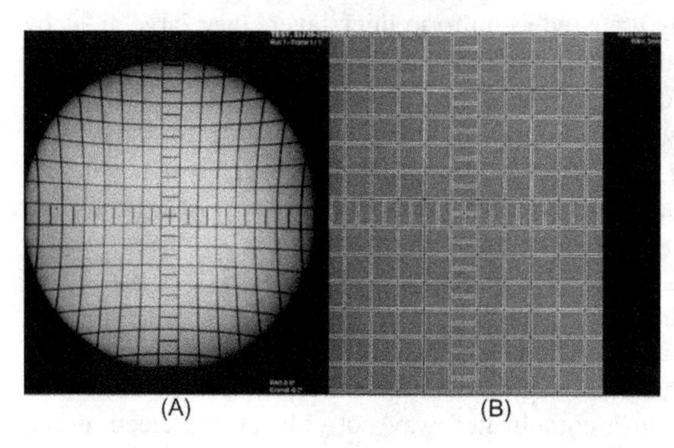

Figure 5.12 Comparison of the distortion in large field image intensifier (A), and solid-state dynamic flat-panel detector (B).

thickness (δ/L). There is an advantage to applying a negative polarity to the front electrode since it avoids damage to the TFT switch on exposure. Positive polarity requires more complicated protection schemes (Figures 5.12 and 5.13).

Many of the degradations associated with the use of combined image intensifier and television camera systems (Figure 5.12) are not present in images obtained

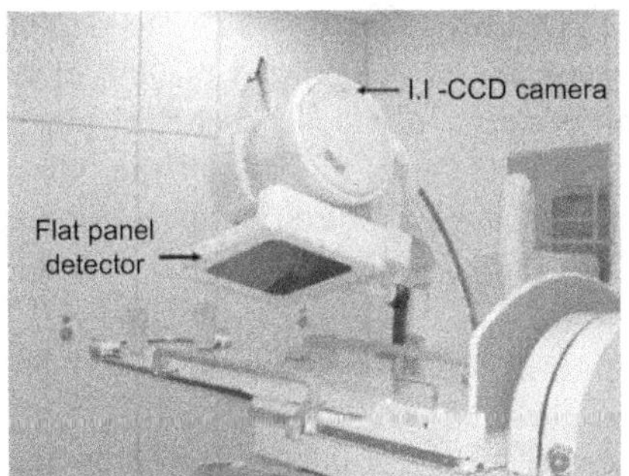

Figure 5.13 This is experimental R&F equipment used by Hitachi in a clinical evaluation of flat-panel technology. Here a 12″ × 16″ (active area) FPD is mounted on the side of a 12″ Image Intensifier Tube (IIT). With this system, either the FPD or the IIT could be rotated into place, facilitating straightforward comparison of images.

with FPD fluoroscopy systems (Figure 5.13). Images obtained with FPD systems also do not exhibit geometric deformation such as the "pincushion" effect and S distortion because the individual DELs (detector element) in the FPD array are manufactured in straight rows and columns. Consistent production techniques and appropriate software calibration ensure excellent uniformity; the distortion that occurs with the use of image intensifier systems is not present in images obtained with FPD systems. In addition, because each DEL is fixed in a constant position, images obtained with FPD systems do not exhibit defocusing effects.

Other advantages of the FPD system include its smaller size, which makes it easier to position during clinical studies, and its solid-state design, which makes it more reliable. FPD systems do not require a television camera to convert the X-ray intensity distribution into an electronic signal; an electronic signal automatically emerges from the image receptor. Moreover, the video signal emerges from the device in a digital format, which reduces electronic noise.

FPD fluoroscopy systems have their own unique limitations. As was previously mentioned, it is difficult to manufacture an FPD array that contains no defective or degraded DELs; if there are too many defective DELs, image quality suffers. Manufacturers of FPD systems often compensate for defective DELs by using software to interpolate values for those defective elements. However, this interpolation may introduce artifacts. Moreover, FPD systems usually are temperature sensitive, and the images may be affected by changes in temperature. FPD detector arrays also are sensitive to mechanical shocks, which can permanently damage the device. Damaged FPD systems can be expensive to replace.

Another limitation of the FPD system is its spatial resolution, which is influenced by the size of its DEL (pitch) and by a process called binning. The pitch is the actual distance between the centers of two adjacent DELs. The best spatial resolution that may be obtained by an FPD is related to the size of the DEL; this spatial resolution is equal to 1 divided by twice the DEL size (in millimeters).

In actual practice, the maximum spatial resolution is about 75–80% of this value because of misalignment of the line-pair pattern with the individual pixels during measurement procedures. By rotating the test pattern to 45° with respect to the rows of the FPD, the spatial resolution lost to misalignment is regained. The typical spatial resolution of most FPD arrays is approximately 2.5–3.2 line pairs per millimeter if the test pattern is placed at 45°.

One may conclude that manufacturers of FPD arrays should reduce the size of the DEL to improve spatial resolution. However, the switching and readout electronics plus the etched data lines on the circuit occupy a portion of each DEL; only a fraction of the total surface area is used to form the image. Some of the X-ray radiation is incident on readout electronic elements and is not used in image formation. The actual fraction of the incident radiation that is available for image formation is called the fill factor.

As the pitch of the DEL decreases, the readout electronics constitute a larger portion of the total surface area, and the efficiency (fill factor) of the DEL drops dramatically. Even normal-sized DELs have an efficiency of only 60–80% for the use of incident X-rays, and smaller-sized DELs are much less efficient. In addition, the amount of radiation incident upon each DEL decreases as its size decreases. For these reasons, although smaller-sized DELs improve the achievable spatial resolution, the images would have more mottle (noise) and would thus require more radiation to reduce this mottle.

Another factor to consider is that large FPD fluoroscopy systems have considerable data rates. For instance, a $40 \times 40\ \text{cm}^2$ FPD system may produce an image composed of 4 million pixels (DELs), an image size of 8 MB, and a data rate as high as 240 MB/s. Large data rates such as these are difficult for electronic systems to handle. To reduce the size of data rates, manufacturers group the data from four DELs together for larger FOVs, a process called binning. Grouping four DELs together reduces the data rate to 25% of the ungrouped rate for large FOVs. Binning has the disadvantage of less spatial resolution because the effective area of each image pixel is four times larger, and it has the advantage of lower data rates and less image mottle than ungrouped DELs.

For smaller FOVs, collimation is used to select only the central portion of the FPD for imaging, which is similar to the process used with image intensifier fluoroscopy systems; thus, information from a smaller anatomic area is spread across the display monitor or magnified. When smaller FOVs are used, the data rate is lower, and binning is no longer required. Unlike image intensifier fluoroscopy systems, the spatial resolution of FPD fluoroscopy systems is the same for all FOVs. For those larger FOVs when binning is employed, the spatial resolution dramatically decreases to 50% of the value without binning; this process is illustrated in Figure 5.8B for a system with a pitch of 200 mm. For FPD systems, there is a dramatic, discrete step change in spatial resolution between small and large FOVs. In addition, for FPD systems, the radiation dose levels to the patient could potentially be the same for all FOVs. The size of pixels (DELs) is the same for all FOVs (provided that binning is not employed); therefore, the amount of X-ray flux on each DEL is the same for all FOVs. Unlike with image intensifier systems, which progressively use more radiation as the FOV decreases, there is no reason to increase the radiation to the image receptor as the FOV is changed. Nevertheless, most vendors of FPD fluoroscopy systems increase the radiation to the image receptor at a rate of approximately 1/FOV. With the selection of smaller FOVs, the dose rates are gradually increased by the FPD system. With smaller FOVs, magnification of the surface area makes the image noise more apparent to the eyes of the observer. In FPD systems, increased radiation for smaller FOVs is used to reduce the optical perception of noise. However, this increase in radiation is substantially less than that used with image intensifier systems. Unlike in image intensifier systems, the thickness of the CsI layer may be increased without substantially degrading spatial resolution of images obtained with FPD systems. For these reasons, FPD fluoroscopy systems are more efficient and tend to require less radiation than image intensifier systems.

Finally, FPD systems have a large operational dynamic range, about 60 times larger than that of image intensifier systems. For this reason, FPD systems do not exhibit flare or veiling glare, which degrade image quality.

Solid-state FPD image receptors generally have better stability, lower radiation dose rates, and improved dynamic range, and they eliminate glare and geometric distortions (e.g., defocusing effects).

The disadvantages of FPD systems include higher costs, lower spatial resolution with very small and very large FOVs, and a different appearance of the displayed image compared with that of images from image intensifier fluoroscopy systems.

5.9 Medical Applications

The medical applications for which flat-panel detectors are being developed and the new opportunities made possible with this technology will be discussed in the context of the procedures currently used in clinical practice. All clinical applications will benefit from the following general features of flat-panel detectors, including compactness, ability to be read out immediately after radiation exposure to verify patient position and appropriate image exposure, ability to permit digital storage and communication within the hospital and beyond, facilitation of computer-aided diagnosis and "second opinions", and perhaps, most importantly, the possibilty improving image quality without increasing patient X-ray exposure due to their enhanced detective quantum efficiency. Another less-recognized benefit is their inherent computer control and that the majority of modern X-ray machines are microprocessor controlled. The computer synchronization of the delivery of the X-ray exposure, the acqusition and readout of the image, and the movement of the X-ray tube and other mechanical devices such as filter holders as well as computer control of the X-ray energy is, thus, possible. Complex imaging procedures, such as dual energy and tomographic data acqusition, improve the busy clinical environment. The following sections will describe the requirements of important clinical imaging tasks and the improvements possible by digital X-ray imaging and flat-panel X-ray detectors.

5.10 Chest Radiography

Flat-panel active matrix X-ray imagers have been configured for chest imaging. The foremost requirements are a very large FOV, a reasonably high spatial resolution (100–200 μm pixels), and a very large dynamic range to accommodate the different penetration of the lungs and mediastinum. Digital image processing can be used to equalize the appearance of the image and, thus, lower X-ray beam will be used in the future.

5.11 Mammography

Mammography is the only projection X-ray imaging modality that attempts to visualize soft tissue contrast and, thus, requires very highly absorbing beams. Film/screen is the current gold standard, but it has a small dynamic range. Therefore, extreme of breast compression to equalize the X-ray path length is needed so that the whole breast can be visualized. Digital mammography is still undergoing development but has potential advantages of increased dynamic range, less breast compression, and the ability to visualize denser breasts. The challenge is to make pixels small enough at an affordable cost. The intrinsically high resolution of a-Se combined with the relative simplicity of the AMA design used for direct conversion suggests that this may be an ideal approach for mammography.

5.12 Fluoroscopy

Perhaps the most demanding potential application for flat-panel imaging systems is in fluoroscopy. Very high patient doses result from lengthy interventional fluoroscopic procedures. During these proceduers, low radiation exposure rates must be used to reduce the total exposure to the patient. This sets a stringent limit on the system performance, since the image quality must still be adequate for visualization of the interventional tools as well as anatomy. Therefore, the imaging system must be X-ray quantum limited, even at extremely low exposure levels, which implies a very low system noise for the AMA. The current technology uses a large vacuum tube devices—an X-ray intensifier. Active matrix panels are more compact, permit better access to patients, and since panel is flat, it is largely free from geometrical distortions characteristic of vacuum tube image intensifier. Thus, quantitative image analysis, registration, and clinical comparison of images from other modalities, 3D construction (e.g., cone beam volume computerized tomography), and use in conjunction with MRI are facilitated [2].

5.13 Future Trends

Flat-panel X-ray detectors are still relatively new. Many advances in system design and improvements in system performance can be expected. As fabrication techniques and device yields improve, more sophisticated switching structures with reduced coupling capacitance, lower leakage currents, smaller physical area, and more robust operating characteristics will continue to be developed. These advances will improve the imaging performance of AMAs until the dominant factor becomes the properties of the X-ray detection medium, even for the most demanding low signal level and high resolution applications such as fluoroscopy and mammography.

The investigation of large area flat-panel sensors presents a large variety of previously unexplored problems in detector physics. How they may be resolved has been discussed. It is to be expected that at the end of this development, an essentially ideal X-ray imaging detector will be possible. The initial investment has already been high, but over time the image quality and labor saving will justify this investment. In time, mass production will eventually reduce the cost.

References

[1] S.O. Kasap, J.B. Frey, G. Belev, O. Tousignant, H. Mani, L. Laperriere, A. Reznik, J. A. Rowlands, Phys. Status Solidi B 246 (2009) 1794.

[2] S.O. Kasap, J.A. Rowlands, Proc. IEEE 90 (2002) 391.

[3] S.O. Kasap, J.A. Rowlands, J. Mater. Sci. Mater. Elect. 11 (2000) 179.

[4] S.O. Kasap, J. Rowlands, B. Fogal, M.Z. Kabir, G. Belev, N. Sidhu, B. Polishchuk, R. E. Johanson, J. Non-Cryst. Solids 299-302 (2002) 988.

[5] D.L. Lee, I.K. Cheung, I. Jeromin, Proc. SPIE 2432 (1995) 237.

[6] D.L. Lee, I.K. Cheung, E.F. Palecki, I. Jeromin, Proc. SPIE 2708 (1996) 511.

[7] U.S. Philips Corporation, USA, 1995, US Patent 5 396 072 (7 May 1995) "X-ray Image Detector".

[8] K. Suzuki, in: I. Kanicki (Ed.), Amorphous and Microcrystalline Semiconductor Devices: Optoelectronic Devices, Artech House, Boston, MA, 1991 (Chapter 3)

[9] J.W. Boag, Phys. Med. Biol. 18 (1973) 3.

[10] I. Brodie, R.A. Gutcheck, Med. Phys. 12 (1985) 362.

[11] D.M. Pai, B.E. Springett, Rev. Mod. Phys. 65 (1993) 163.

[12] S.O. Kasap, V. Aiyah, B. Polischuk, A. Baillie, J. Appl. Phys. 83 (1998) 2879.

[13] S.O. Kasap, in: A.S. Diamond, D.S. Weiss (Eds.), Handbook of Imaging Materials, second ed., Marcel Dekker, Inc., New York, NY, 2002, p. 329; and references therein.

[14] W. Que, J.A. Rowlands, Med. Phys. 22 (1995) 2029.

[15] W. Knupfer, Nuc. Phys. B 78 (1999) 610.

6 Effects of Charge Carrier Trapping on Detector Performance

6.1 Introduction to the Trap Level Spectroscopy

Parameter $\mu\tau F$ represents the mean distance drifted by the carrier before it is trapped (or disappears by recombination); this distance is called the schubweg. Since we need to collect most of the charges, both electrons and holes, we need to ensure that the electron and hole schubwegs are both much longer than the thickness of the photoconductive layer, that is, $\mu\tau F \gg L$ for both electrons and holes. If the photoconductor layer is made thicker to capture more of the radiation (toward increasing A_Q), the $\mu\tau F \gg L$ condition would eventually be lost, and charge collection efficiency would start limiting the sensitivity. The charge collection efficiency depends on the photoconductor properties.

The drift mobility × lifetime product, $\mu\tau$ is normally called the range of the carrier, that is, its schubweg per unit field. Table 6.1 compares the carrier ranges among various large area X-ray photoconductors. One of the distinct advantages of a-Se is the fact that both electrons and holes possess reasonable ranges, which allows both electrons and holes to be collected upon their photogeneration. It is well known that a-Se exhibits typical polymeric glass properties.

One of the most important parameters which determine the performance of many modern devices based on amorphous semiconductors is the drift mobility-lifetime product, $\mu\tau$. There has been much interest in determination of charge-carrier ranges in amorphous semiconductors by various measurement techniques. Although the mobility, μ, can be measured by the conventional time-of-flight (TOF) transient-photoconductivity technique, the determination of the lifetime, τ, is often complicated by both experimental and theoretical limitations. The present paragraph provides an overview of xerographic measurements as a tool for studying the electrical properties of amorphous semiconductors.

First, details of the experimental set-up are discussed. Thereafter, the analysis and interpretation of dark discharge, the first-cycle residual potential, cycled-up saturated residual potential are considered. It is shown that from such measurements the charge-carrier lifetime, τ, the range of the carriers, $\mu\tau$, and the integrated concentration of deep traps in the mobility gap can be readily and accurately determined. Xerographic measurements on Se-rich amorphous photoconductors have indicated the presence of relatively narrow distribution of deep hole traps with integrated density of about 10^{13} cm^{-3}. These states are located at ~0.85 eV from the

Medical Imaging Technology. DOI: http://dx.doi.org/10.1016/B978-0-12-417021-6.00006-X

Table 6.1 Some Typical Properties of Selected X-Ray Photoconductors for Large Area Applications

Photoconductor State Preparation	δ at 20 keV δ at 60 keV	E_g eV	$W_{\pm}$ eV	Electron $\mu_e\tau_e$ (cm^2/V)	Hole $\mu_h\tau_h$ (cm^2/V)
Stabilized a-Se Amorphous Vacuum deposition	49 μm 998 μm	2.2	45 at 10 V/μm 20 at 30 V/μm	3×10^{-7}–10^{-5}	10^{-6}–6×10^{-5}
HgI_2 Polycrystalline PVD	32 μm 252 μm	2.1	5	10^{-5}–10^{-3}	10^{-6}–10^{-5}
HgI_2 Polycrystalline SP	32 μm 252 μm	2.1	5	10^{-6}–10^{-5}	$\sim10^{-7}$
$Cd_{.95}Zn_{.05}Te$ Polycrystalline Vacuum deposition (sublimination)	80 μm 250 μm	1.7	5	$\sim2\times10^{-4}$	$\sim3\times10^{-6}$
PbI_2, Polycrystalline Normally PVD	28 μm 259 μm	2.3	5	7×10^{-8}	$\sim2\times10^{-6}$
PbO, Polycrystalline Vacuum deposition	12 μm 218 μm	1.9	8–20	5×10^{-7}	small

δ is the attenuation depth at the shown photon energy, either 20 keV or 60 keV. $\mu\tau$ represents the carrier range. SP refers to screen printing, and PVD to physical vapor deposition.
Source: From various references (see Ref. [48] for details).

valence band. A good correlation was observed between residual potential and the hole range, in agreement with the simple Warter expression. The capture radius is estimated to be $r_c = 2 - 3$ Å. Since r_c for pure a-Se and a-As_xSe_{1-x} is comparable to the Se–Se interatomic bond length in a-Se, it can be suggested that deep hole trapping centers in these chalcogenide semiconductors are neutral-looking defects, possibly of intimate valence-alternation pair (IVAP) in nature. The absence of any electron spin resonance signal (ESR) at room temperature seemed to be a strong argument in favor of this suggestion. Finally, photoinduced effects on xerographic parameters are discussed. It has been shown that photoexcitation of a-As_xSe_{1-x} amorphous films with band-gap light alters deep hole and electron states. During room-temperature annealing photosensitized states relax to equilibrium. Recovery process becomes slower with increasing As content. Qualitative explanation of the observed behavior may be based on associating the deep states with $C_3^+$and C_1^- IVAP centers.

Historically, selenium has played an extremely important role in physics and technology, and they continue to do so. It is well known that at the very beginning crystalline selenium was used in photocells, photodiodes, and rectifiers. On the other hand, investigation of amorphous semiconductors is one of the most attractive disciplines of condensed matter physics. These investigations are strongly stimulated by both scientific and technological factors. Amorphous semiconductors have a wide range of applications but the first and most impressive was in the xerography. Amorphous selenium (a-Se) is one of the earliest and most highly developed photoconductors that served the photocopying industry for over three decades. Pure selenium can be easily coated in a thick film (50–500 μm) form onto large area suitable substrates by simple vacuum deposition at a fast rate (e.g., 1–5 μm/min). Later various alloys of a-Se (e.g., Se–Te, Se–As) were also used as photoreceptors.

Before the development of electrographic engines, other inorganic photoreceptors like ZnO-type (ZnO suspended in a polymer matrix) were used in special paper-copiers. Photoreceptors based on ZnO have serious disadvantage: they have short "lifetime" because of charge trapping problems. CdS_nSe_m photoreceptors have been developed and used in some Japanese copiers.

It seemed that a new inorganic photoreceptor developed in recent decades, namely hydrogenated amorphous silicon (a-Si:H), may play significant role in the future. The advantages of this material include:

a. excellent photosensitivity over a broad spectral range (450–750 μm); even more, the photosensitivity can be extended into the infrared by alloying with Ge
b. high surface hardness
c. mechanical strength
d. thermal stability.

This material permits to obtain 10^6 copies or higher and have 10 times longer "lifetime" than a-Se. At the same time, there is a problem, caused by manufacturing cost—plasma discharge decomposition of silane requires several hours and corresponding complicate equipment.

Although organic photoconductors currently dominate the xerographic photoreceptor industry, one nonetheless still finds amorphous inorganic semiconductor photoreceptors in a limited number of copying machines.

There are however new imaging technologies in which a-Se alloys based photoconductors cannot be simply replaced by organic photoconductors due to a number of unique features. These features are used in either commercial products (HARPICON, Thoravision) or in prototype units (flat panel digital X-ray image detectors) and in a large area X-ray vidicon [1].

The performance of many amorphous semiconductor devices is related to the range of the charge carriers. This parameter is defined as the drift mobility (μ) and lifetime (τ) product, $\mu\tau$. The determination of $\mu\tau$ is of fundamental importance in the characterization of amorphous semiconductors. Reasonably, there has been much interest in determination of charge-carrier ranges in amorphous semiconductors by various measurement techniques. Xerographic experiments have been used since they provide clear measurements of a measurable surface potential, termed the residual potential, due to trapped charges in the bulk.

The general consensus in the literature is that $\mu\tau$ can be readily determined via simple xerographic measurements. The simplest theoretical model, based on range limitation and weak trapping ($V_R \ll V_0$), relates V_R to $\mu\tau$ via the Warter equation [2]

$$V_R = L^2/2\mu\tau \tag{6.1}$$

where L is the sample thickness. Somewhat latter, a more general description of the deep trapping phenomenon was made by Kanazawa and Batra [3]. They have solved the charge transport, continuity, detailed balance and Gauss's equations and obtained a universal curve. This curve allows $\mu\tau$ being determined from the measurement of the residual potential:

$$V_R \approx (L^2/2\mu\tau V_0)[-\ln(2V_R/V_0)] \tag{6.2}$$

where V_0 is the charging voltage. Equation (6.1) has been found [4] to predict the actual $\mu\tau$ product remarkably well. The paper [5] identifies and critically examines the theoretical problems involved in the determination of $\mu\tau$ from xerographic measurements. It is found [5] that a single xerographic measurement in general cannot be used to accurately determine the $\mu\tau$ product. Xerographic measurements can be combined with conventional TOF [6,7] and interrupted-field time-of-flight (IFTOF) [8] techniques to check the measurement of the carrier range, and to calculate the deep trap capture radius. There is no other technique in which μ and τ are measured independently; and can even be measured as a function of sample composition across the sample thickness. Although the mobility, μ, can be measured by the conventional TOF transient-photoconductivity technique, the determination of the lifetime, τ, is often complicated by both experimental and theoretical limitations. The early work involving drift mobility measurements was pioneered by Spear who successfully applied the technique to high-resistivity solids in the 1960s [6]. Nearly all recent TOF experiments have been based on the conventional principle of applying bias voltage, injecting charge carriers by photoexcitation, and then measuring the resulting transient photocurrent generated in the external circuit. There is a wealth of information in the literature on the interpretation of the photocurrent waveform in terms of the interaction of the photoinjected carriers with various traps in the sample. In the case of amorphous semiconductors, the TOF photocurrent provides a means of obtaining the energy distribution of the localized states in the mobility gap [7]. There are several variations of the conventional TOF measurements. They involve the following:

a. delayed bias (or advanced photoexcitation) for studying transport [9];
b. delayed photogeneration for studying bulk space–charge evolution [10];
c. double photoexcitation for recombination studies [11].

In general, standard TOF measurement cannot be used directly to determine charge-carrier lifetime τ. When the lifetime is less than the transit time, $t_0 = L^2/\mu V$, where V is the applied voltage, then the transport is range limited and the photocurrent decays rapidly without clear indication of transit time. Therefore, as the applied voltage is reduced the transit time becomes comparable to the deep trapping time and the transport approaches the range-limited response regime. In such a case, the carrier lifetime can be determined via Hecht analysis of the TOF signal [12–17]. The technique becomes much more difficult when the quantum efficiency is strongly field dependent, or when the applied field is comparable to any built-in field. Nevertheless, the transient-photoconductivity technique can still be used to determine the carrier lifetime by operating the TOF method in the interrupted field mode [18]. The IFTOF measurement, in essence, involves interrupting the transit of photoinjected charge carriers during their flight across the specimen in the conventional TOF measurement. The interruption is achieved by removing the applied field at time $t = T_1$. After an interruption period Δt, the field is reapplied at time $t = T_2$ to collect the remaining carriers. The fractional change I in the recovered photocurrent is related to the lifetime via

$$I = i_2/i_1 = i(T_2)/i(T_1) = \exp(-\Delta t/\tau) \tag{6.3}$$

where $i(t)$ is the instantaneous photocurrent and τ is the mean lifetime.

Charge transport and trapping in a-Se and its alloys has been a subject of much interest [7,13,19–26]. The nature of charge transport in pure and alloyed a-Se has been studied by the TOF transient-photoconductivity technique. Both hole and electrons are mobile in a-Se (μ_h is larger than μ_e by an order of magnitude). The drift mobility at low temperatures in these materials is thermally activated. The whole conduction mechanism in a-Se is shallow-trap-limited extended-state transport. Various impurities and alloying elements have drastic effect on the nature of charge transport in a-Se [19–26].

6.2 Technique

When a high voltage of several kilovolts is applied to a corona emitter, the field near the emitter exceeds the threshold field for air breakdown. Under this condition, the molecules near the emitter become ionized. The ions may carry positive or negative charge. This depends on whether the emitter is at a positive or negative voltage. If insolating material which is supported by earthed metal is placed close to the corona emitter, some of the generated ions migrate to the floating surface of the sample and increase or decrease the surface charge density.

To improve the charging performance, the grounded electrodes are placed around the corona emitter. The structure is called the corona housing and helps direct the ion current and holds the corona emitter. Screen grid is inserted between the corona emitter and the sample to provide better control of charging. This is the corona grid. Several variables are known which determine the efficiency of sample charging. Among these the most important are:

a. The uniformity of the charge deposited on the sample surface.
b. The amount of corona current which reaches the sample surface.

Reasonably, the assessment of electrostatic characteristics of the sample examined depends on its particular applications. As for electrophotography, the following may be listed: measuring the charge acceptance (or the maximum surface charging potential), dark decay of the surface potential of the sample charged in darkness, light decay or photoinduced discharge (PID), and so-called residual potential (in other words, the potential which remains on the sample surface after photodischarge).

There are two, more or less exploited types of corona devices—a pin corona discharge device and a wire corona discharge device. These devices are well described by Vaezi-Nejad [27].

A pin corona discharge corona device consists of a stainless steel rod with a sharp conical tip as the corona emitter, a brass hollow cylinder as the corona case, and a Perspex cup as the corona emitter holder.

Corona devices of this type have been used in the past (see, e.g., [28]). Latter becomes clear that the electrostatic charging performance of pin corona is unsatisfactory.

Most important component of the apparatus for measurement of corona current distribution is the current sensing probe. This is a small aluminum disk (area $= 1.77\ mm^2$) placed in the center of a relatively large plate (area $= 4.2 \times 2.6\ cm^2$). The detecting disk is electrically insulated from the plate by a 0.75 mm air gap. As a disadvantage of a pin corona device it should be mentioned that the current distribution is not uniform.

An alternative to pin corona is a single wire corona. A wire corona basically consists of a hollow stainless steel cylinder (inner diameter $= 3-5$ cm, length $= 10.7$ cm, and thickness $= 0.16$ cm) as the corona case with a $5.7\ cm \times 3.0\ cm$ window. Two Perspex slabs of dimensions $5.5 \times 3.4 \times 1.25\ cm^3$ were used to hold the corona wire. The primary requirement for a corona wire emitter is that it should produce ions by means of a corona discharge at a reasonable voltage which is typically 3–7 kV. In order to achieve high electric field, the wire diameter must be small (less than 100 μm). As the material for corona emitter, tungsten is particularly suitable because of its resistance to the harsh environment created by the corona discharge such as ultraviolet light and various nitrogen–oxygen compounds produced in the discharge. Another advantage of this material is that it is mechanically strong to resist breakage caused by the stress of stringing, handling, and cleaning. The corona onset voltage of 1 kV was easily achievable. It is possible to improve a single corona (this type of corona device we have considered earlier) charging characteristics by adding an identical tungsten wire to the device thus forming a double wire corona discharge. The wires were 3.7 cm long and spaced by 0.5 cm one another.

In an attempt to find most efficient corona discharge device for xerographic spectroscopic purposes, it seems necessary, as Vaezi-Nejad did [29], examine various device configurations utilizing sharp pin/pins and wire/wires as the corona emitter.

Experimental results obtained [28,30] illustrate clearly that devices based on thin wires provide a more uniform charge distribution and thus may be recommended for xerographic TOF spectroscopy. In order to control the sample initial voltage, a biased grid is inserted between the corona emitter and the floating surface of the sample.

6.3 Measurement Technique

After charging the sample, it is necessary to measure exactly the surface potential at various stages. These are the following:

1. The initial voltage of the sample for calculation of drift mobility.
2. The sample voltage after photoexcitation. One can estimate the total charge injected into the sample. This can be made subtracting the sample voltage (after photoexcitation) from the initial voltage.
3. The final value of the sample voltage which is measured after photoexcitation. This is known in the literature as the residual voltage and it is due to deeply trapped photoinjected carriers when they transit across the sample. If the sample is repeatedly charged–discharged, the residual voltage builds up with the number of cycles and saturates. In this case, we deal with the saturated residual voltage and can extract information about the deep traps.

From these measurements important information can be extracted. In last decades, the xerographic probe technique becomes a very popular and unique means to characterize electronic gap states. In particular, a map of states near mid-gap is determined by time-resolved analysis of the xerographic surface potential [31,32].

An optimal photoreceptor design will require, among many other factors, high charge acceptance, slow dark discharge, low first and cycle-up (saturated) residual voltages, and long carrier ranges ($\mu\tau$). The later factor has been addressed earlier.

In general, the dynamic behavior of electrophotographic potentials developed at two important stages in the xerographic cycle: immediately after charging in the dark (dark decay) and then immediately after photodischarge (i.e., residual decay). Both are extraordinarily sensitive to bulk space−charge fluctuations. In fact, $10^{12}\ cm^{-3}$ uniformly trapped electrons can, in a typical sample film, give rise to several volts of surface potential which is easily measured. From xerographic measurements, especially from time-resolved analyses, information can be extracted about electronic gap states which determined the photoelectronic behavior of amorphous chalcogenides.

There are essentially three important types of xerographic behavior. They are generally termed:

a. the dark discharge
b. first-cycle residual
c. the cycle-up residual voltage.

All must be considered in evaluating the electrophotographic properties, e.g., of a-Se and its alloys. It should be stressed that these three parameters are extremely informative in the sense to map band-gap states.

In the sensitization-exposure part of the xerographic cycle, a photoreceptor film mounted on a grounded substrate is charged to a voltage V_0 by corona then completely photodischarged by exposure to strongly absorbed light (absorption depth $\delta \ll L$; L is the sample thickness). Both the top corona contact and the substrate contact should be blocking. The resulting capacitor-like structure contains a uniform field after charging in the dark. Illumination with strongly absorbed light creates a thin sheet of electron−hole pairs which separate under action of field. Depending on charging polarity one sign of carrier acts to neutralize charge on the top surface while the opposite sign of carrier drift through the bulk toward the grounded surface. As a result of this carrier displacement, the surface voltage decays in time: this is called PID. Ideally, the surface voltage would decay to zero. What really happened is that some fraction of carriers becomes deeply trapped during transit through the bulk. Reasonably, at the end of the illumination, there is a measurable surface potential termed the residual potential, V_R. Space−charge neutrality is finally reestablished by thermally stimulated process but on a longer time scale.

It should be noted that there is important information yielded at each stage of the xerographic cycle. This information consists of the following:

1. By time resolving the charge delivered to an amorphous film from a corona, one can discern injection phenomena and measure dielectric parameters.
2. By measuring the temperature-dependent dark decay of surface voltage, one can determine the energy distribution of thermal generation centers. This later technique or the measurements of the dark discharge depletion time, which is the time, required for the bulk to generate some fixed quantity of charge under isothermal condition, one can estimate the position of the Fermi energy. We consider the details of depletion discharge latter.
3. Analysis of PID can be used to determine the parameters which characterize photogeneration and transport processes.

4. From analysis of residual buildup during repetitive cycling and from analysis of isothermal decay of xerographic residual after cycling ceases, it is possible to map the spectral distribution of bulk traps in an amorphous film.

Corona-mode xerographic measurements were carried out using a reciprocating sample stage. In operation, the sample is first passed under a corona-charging device, the corotron, which can be set to deposit either positive or negative ions on the sample surface. The charging circuitry is such that constant current can be supplied to the sample, simplifying the residual voltage cycle-up and in some cases xerographic TOF data. After charging, the sample is moved to a measuring station and the surface voltage is determined using a transparent, capacitively coupled electrometer probe. The time interval between termination of the charging and the onset of surface voltage measurement is a few tenths of seconds. Either the dark decay or the photoinduced-discharge characteristic (PIDC) following step or flash illumination with strongly absorbed light, can be measured. After discharge, decay of residual potential can be time resolved. The contacts are blocking in the sense that they allow the samples to charge capacitively in the dark for both positive and negative corona.

Dark discharge rate must be sufficiently low to maintain ample amount of charge on the photoreceptor during the exposure and development steps. A high dark decay rate will limit the available contrast potential. The residual potential remaining after the xerographic cycle must be small enough not to impair the quality of the electrostatic image in the next cycle.

In the case of a-Se, these xerographic properties have been extensively studied by numerous authors. In addition to the magnitude of the saturated residual voltage, the rate of decay and the temperature dependence of the cycle-up residual potential are important considerations, since they determine the time required for the photoreceptor to retain its first-cycle xerographic properties.

The simplest procedure for xerographic measurements is the following. The rotating photoreceptor drum is charged by a corotron device. The surface potential is measured at this position, and the photoreceptor is then exposed to a controlled wavelength and intensity illumination at the next station, following which its surface potential is measured again. In some systems, the surface potential is also monitored during exposure via a transparent electrometer probe to study the PID characteristics. Normally, the charging voltage, speed of rotation, and exposure parameters such as energy and wavelength are user-adjustable.

The entire sequence of charging, photodischarge, and residual can be continuously recycled. As a consequence, one can observe the progressive, stepwise buildup of residual voltage. Over many cycles, the cycle-up residual potential should be small, to avoid deterioration in the copy quality after many cycles.

The residual potential can be related to the space–charge density. For the case of uniform bulk space–charge of density ρ_0, we may write [31,32] the simplified expression for residual potential V_R

$$V_R = \rho_0 L^2 / 2\varepsilon_0\varepsilon \tag{6.4}$$

or equivalently

$$V_R = NeL_2/2\varepsilon_0\varepsilon \tag{6.5}$$

where L is the sample thickness, ε_0 is the relative dielectric constant, ε the free-space permittivity, N is the number per unit volume of electronic charges e. A relatively small density of surface or bulk-trapped charges on or in typical xerographic film can give rise to an appreciable surface voltage [31]. The $\mu\tau$ product where μ is the drift mobility and τ the bulk deep trapping lifetime can be determined from first-cycle residual potential of a well-rested sample of thickness L. In the weak trapping limit (when the residual V_R is much less than the charging voltage V_0), Kanazawa and Batra [3] derive the expression (6.2). The physically plausible expression derived by taking the residual to be that voltage for which the carrier range is nominally half the sample thickness

$$V_R/V_0 \approx 0.5L^2/\mu\tau V_0 \tag{6.6}$$

In the strong trapping limit $V_R(V_r \sim V_0)$ Kanazawa and Batra [3] derive

$$V_R/V_0 = 1 - (\mu\tau V_0/L^2)(\ln 2) \tag{6.7}$$

which differs only slightly from the expression based on the above physical definition [33] of residual

$$V_R/V_0 = 1 - (\mu V_0\tau/L^2) \tag{6.8}$$

It should be noted here that bulk deep trapping lifetimes computed from first-cycle residuals are in agreement with lifetimes measured in the TOF mode under range-limited conditions [34]. For example [31], in an a-Se specimen with $L = 48\ \mu m$ charged to 200 V, hole and electron residuals at $T = 295$ K were found to be 1.8 and 44 V, respectively. The measured drift mobility, lifetimes, and $\mu\tau$ products, for electrons and holes, respectively, are 4.9×10^{-3} cm^2/V s, 5×10^{-5} s, 2.45×10^{-7} cm^2/V, and 0.16 cm^2/V s, 4.4×10^{-5} s, 7.04×10^{-6} cm^2/V. The $\mu\tau$ product computed using Eq. (6.5) is 2.62×10^{-7} cm^2/V for electrons and 6.4×10^{-6} cm^2/V for holes.

6.4 Dark Discharge in a-Se

The surface potential of charged photoreceptor will decay even in the dark. The dark discharge rate depends on substrate temperature.

In principle, thermal generation of carriers in the bulk, hole, and electron transport and interfacial injection—all these factors can be the cause of dark decay. The simplest picture of dark decay is one in which electron–hole pairs are thermally generated in the bulk at a rate G_B ($cm^{-3}\ s^{-1}$) and swept put rapidly compared with the generation rate [35]. The superimposed effect of injection is expressed via a surface charge generation [36] rate J_S($cm^{-2}\ s^{-1}$).The time-dependent field E can be obtained by setting the sum of the displacement and conduction currents equal to zero:

$$\frac{dE}{dt} = e(J_s + G_g L)/\varepsilon \tag{6.9}$$

To distinguish bulk from surface contribution to it dark decay, a series of measurements on samples of varying thickness are required.

The dark decay in pure and alloyed a-Se films may be caused by all of the above-mentioned factors: substrate injection, bulk thermal generation, and depletion. It has been found that that the depletion-discharge mechanism dominates. At present, this mechanism is generally accepted [37–45].

Experimental studies of dark discharge performed on various Se-based alloys clearly illustrate that instead of the type of behavior predicted by Eq. (6.8), dark decay is characterized by two distinct zones of time dependence separated by an abrupt transition.

This feature can be completely accounted for by a model which presumes that only one sign of carrier is mobile. During dark decay, the mobile carrier is depleted (in other words swept out), leaving behind a space–charge of opposite sign. Therefore, the dark decay is called a depletion discharge.

In the following, we consider the depletion-discharge model which describes a dark discharge with only one sign of thermally generated charge carrier is mobile on the time scale of experiment. As thermal generation and sweep out of the mobile carrier proceeds, the bulk develops deeply trapped space–charge $\rho(x)$ of opposite sign. The time-dependent surface voltage is computed at any time by integrating the instantaneous field across the sample thickness, taking into account of all sources. This procedure directly allows one to distinguish the bulk charge generation process from surface charge losses.

The instantaneous surface potential is given by [38]:

$$V = \int_0^X dx\, E(x) = \frac{1}{\varepsilon}\int_0^X dx(\sigma - \int_0^L dx\, \rho(x) \tag{6.10}$$

where σ is the surface charge density initially deposited, minus any loss by injection up to instant of measurement. Note that X is the depth (measured from the top surface) at which $E(x)$ is zero and is equal to or less than thickness L. At $X < L$, the surface charge density and X are related by:

$$\sigma = \int_0^X \rho(x) dx \tag{6.11}$$

If depletion is a spatially homogeneous process, Eqs (6.9) and (6.10) simplify and the instantaneous surface voltage is then given by:

$$V = (\sigma X/\varepsilon) - (\rho X^2/2\varepsilon) \tag{6.12}$$

where X and ρ are time dependent.

The process seems to proceed under following scenario. Initially, when the bulk charge is less than the surface charge ($\sigma > \rho L$), then X equals L and is time independent. Latter, when X is less than L, X equals σ/ρ and is therefore time dependent. As a result, with increasing ρ the parameter X decreases. Between these zones exists a demarcation defined by the condition

$$\rho L = \sigma \tag{6.13}$$

which occurs when $t = t_d$, where t_d is called the depletion time. The explanation of physical process at depletion discharge is, at the first sight, simple. As charge is uniformly depleted from the bulk of the sample, ρ increases in time until the depleted charge equals the surface charge (Eq. 6.12). Further depletion now begins to reduce the depleted volume by reducing the dimension $X = \sigma/\rho < L$. As X shifts toward $X = 0$, the region between $x = X$ and $x = L$ regains its space–charge neutrality. The dark discharge rate $\mathrm{d}V/\mathrm{d}t$ and the surface voltage each exhibit different behavior in the two respective time zones separated by t_d. In zone 1, when $t < t_d$, one obtains (at condition that there is no surface charge loss)

$$V = \frac{\sigma L}{\varepsilon} - \frac{\rho L^2}{2\varepsilon} \tag{6.14}$$

and

$$\frac{\mathrm{d}V}{\mathrm{d}t} = -\frac{L^2}{2\varepsilon}\frac{\mathrm{d}\rho}{\mathrm{d}t} \tag{6.15}$$

In zone 2, when $t > t_d$, we obtain

$$V = \sigma^2/2\varepsilon\rho \tag{6.16}$$

and

$$\frac{\mathrm{d}V}{\mathrm{d}t} = -\frac{\sigma^2}{2\varepsilon\rho^2}\frac{\mathrm{d}\rho}{\mathrm{d}t} \tag{6.17}$$

From Eq. (6.13), it is clear that when $t = 0$, $V = V_0 = \sigma L/\varepsilon$ and at t_d when $\rho L = \sigma$,

$$V = \sigma L/2\varepsilon = V_0/2 \tag{6.18}$$

Therefore, the voltage is reduced to half its initial value at the depletion time. Equations (6.14) and (6.16) are modified when surface loss by injection is included by the addition of a surface term $(\mathrm{d}V/\mathrm{d}t)_s$, given by

$$\left(\frac{\mathrm{d}V}{\mathrm{d}t}\right)_s = \frac{L}{\varepsilon}\frac{\mathrm{d}\sigma}{\mathrm{d}t} \text{ in zone 1} \tag{6.19}$$

and

$$\left(\frac{\mathrm{d}V}{\mathrm{d}t}\right)_s = \frac{\sigma}{\varepsilon\rho}\frac{\mathrm{d}\sigma}{\mathrm{d}t} \text{ in zone 2} \tag{6.20}$$

The boundary between the zones is defined by condition $\rho(t_d)L = \sigma(t_d)$. It follows that

$$V(t_d) = \frac{\sigma(t_d)L}{2\varepsilon} = \frac{\sigma(t_d)}{\sigma(0)}\frac{V(0)}{2} \tag{6.21}$$

where $\sigma(0)$ and $V(0)$ are the values at $t = 0$. It should be stressed that the extent to which the surface voltage at t_d has decayed beyond $V_0/2$ thus provides a quantitative measure of the surface loss in specimen.

One representation of $\rho(t)$ is based on the assumption that emission is a first-order rate process. The latter applies to a system containing N_0 (cm^{-3}) discrete emission centers displaced by energy E_0 from the transport state. For this system $\rho(t) \sim t$ because

$$\rho(t) = en = eN_0[1 - \exp(-Rt)] \tag{6.22}$$

where

$$R = \nu \exp(-E_0/kT) \tag{6.23}$$

and

$$\rho(t) \ll eN_0, \quad Rt \ll 1 \tag{6.24}$$

thus

$$\rho(t) \approx eN_0Rt \tag{6.25}$$

From Eqs. (6.12), (6.22), and (6.23), it follows that for emission from a discrete center,

$$t_d = (2V_d)\varepsilon/eN_0L^2\nu \exp(-E_0/kT) \tag{6.26}$$

where we abbreviate $V_d = V(t_d)$ so that in this case a plot of log t_d versus log$(2V_d)$ should be linear. Any curvature when the log of t_d is plotted against log V_0 (V_0 = surface voltage at $t = 0$) is then a measure of surface loss by injection. Here, we note that the scaling of t_d with charging voltage (and the shapes of the associated depletion curves) remained invariant over the entire Cl doping concentration series and in thick films of undoped a-Se. Such a behavior are observed also in Se-rich amorphous alloys and even in pure selenium (in the latter case only in thick, ~50 μm, films charged to relatively low voltages at which surface loss was negligible).

In series of elegant experiments, Abkowitz and Markovics [46] demonstrate that the broadening of shallow transport interactive states identified in TOF experiments (the dispersion of TOF transients is increased with Te concentration in Se:Te alloys compared to pure Se occur in parallel with a similar process operating on the deep gap states which control the dark decay.

The rate at which bulk space–charge is uniformly generated may be represented as

$$\rho(t) = at^p, \quad p < 1 \tag{6.27}$$

This algebraic time dependence is suggested by the analytical modeling of simple cases. Let us consider exponential distribution of emission centers given by [46]:

$$N = N_0 \exp(-[E - E_0]/W) \tag{6.28}$$

which is N_0 at the reference energy E_0, and is also cut off by the Fermi energy for holes (i.e., states extending beyond E_F are hole traps rather than emission centers).

It follows that

$$\rho(t) = eN_0(\nu t)^{kT/W} \tag{6.29}$$

Here it is assumed that the emission rate prefactor v does not depend on energy and there is no retrapping. Substitution Eq. (6.25) into Eqs. (6.14) and (6.16) yields

$$\frac{dV}{dt} = -\frac{L^2}{2\varepsilon} apt^{p-1}, \quad t < t_d \tag{6.30}$$

and

$$\frac{dV}{dt} = \frac{\varepsilon}{2a}(V_0/L)^2 pt^{-p-1}, \quad t > t_d \tag{6.31}$$

We can predict now the following distinctive characteristic for depletion discharge:

a. There is an abrupt change in the dark decay rate ($\sim t^{2p}$) at $t = t_d$. The physical origin of t_d: it is the time required by the bulk to generate isothermally a quantity of charge equal to the surface charge. This parameter (t_d) provides a quantitative measure of thermal generation.

b. The sum of the slopes of two linear segments corresponding to $t < t_d$ and $t > t_d$ equals -2.

c. At $t = t_d$,

$$Lat^p{}_d = \sigma \tag{6.32}$$

or

$$t_d = (\sigma/aL)^{1/p} = (2V_d\varepsilon/\sigma L^2)^{1/p} \tag{6.33}$$

d. When a series of films differing in thickness L are charged to the same initial field, then a transition from quadratic to thickness-independent behavior will be observed in the dark decay rate at t_d.

In the depletion-discharge model, both the shapes of the dark decay and the scaling of the depletion time t_d with the depletion voltage V_d (half the charging voltage in the absence of surface loss) and specimen thickness is completely specified by the p parameter. The form taken by the phenomenological depletion model resembles a key feature predicted in the model proposed by Scher and Montroll [47] for dispersive transport. In the latter model, it is the dispersion parameter α which governs a similar scaling law for the transit time.

It is observed that under fixed conditions the depletion time decreases exponentially with Te concentration while the p parameter decreases linearly over the same composition range.

The above simple phenomenological model which assumes a power law time-dependent development of spatially uniform negative charge density $\rho(t) = at^p$ can account for all features of the dynamic behavior of the surface potential. Thus when plotted on a log–log scale, the (dV/dt) versus t curve for a wide range of alloy films was found to exhibit a well-defined kink precisely at time t_d when the negative bulk-trapped space–charge became equal to the surface charge initially deposited. Depletion time t_d provides a measure of the time required by the bulk to generate an experimentally determined quantity of charge. Both the shape of the decay of (dV/dt) versus t on either side of t_d and the scaling of the depletion time t_d with surface charge σ and sample thickness L, $t_d = (\sigma/aL)^{1/p}$, respectively, provide independent means for measuring parameter p. If the build up in the bulk negative charge density is spatially uniform (as we assume), the internal electric field falls linearly with distance from the top surface. At a certain time t_d, the electric field at the grounded end of the sample becomes zero. From that time onward the field will be zero at a distance $X(t) < L$, the sample thickness, and consequently there will be a neutral region from X to L inasmuch as holes generated in $0 < x < X$ and arriving into $X < x < L$ will not be swept out. The shrinkage of the depleted volume with time $t > t_d$ means that t_d marks a functional change in the dark decay rate and therefore is readily obtainable from dark discharge experiments.

Analysis performed for a range of Se-based alloys reveals that the parameter p and the depletion time t_d scale systematically with alloy composition when experimental conditions remained fixed. Parameter t_d and p decrease with an increase of the concentration of arsenic, antimony, or tellurium in the alloy [38,48–50]. The quantity p in the model for dark depletion discharge plays a role analogous to the dispersion parameter α in the theory of dispersive transport [47] of photoinjected carriers. The parameter p in the depletion model has been related to the width of deep gap thermal emission centers.

As it was mentioned earlier, under low charging voltages the depletion time indicates the time required for the surface potential to decay to half its original value. Under high charging voltages, however, field-enhanced emission from the deep mobility gap centers also plays an important role, and the surface potential initially decays at a much faster rate so that at the depletion time the surface potential is in fact less than half the initial value. Figure 6.1 shows the dependence of the depletion time t_d and the half-time $t_{1/2}$ on the charging voltage V_0, where it can be seen that at the highest charging voltages there is no improvement in $t_{1/2}$ with further increase in the charging voltage V_0. Inasmuch as the dark decay in Se-based alloys is a bulk process, the rate of discharge increases with the square of thickness $dV/dt \sim L^2$ and can be reduced only by using thin layers. The latter concept leads naturally to the design of multilayer photoreceptor structures.

It should be stressed that with thick films, a good blocking contact between a-Se and the preoxidized aluminum substrate, the depletion mechanism dominates the discharge process. There are several reasons for the experimentally observed fact that amorphous selenium possess good dark decay characteristics. These are listed below:

1. There are not many deep localized states in the mobility gap of pure amorphous selenium. Their concentration is low (less than $10^{12} - 10^{13}$ cm^{-3}).
2. The energy location of these localized states is deep ($\geq$0.85 eV) in the mobility gap. Therefore the thermal generation process of holes and/or electrons from these centers is slow.
3. Injection from the substrate can be reduced substantially by using oxidized Al substrate.

The origin of the deep localized states that control the dark decay has been attributed to structural native thermodynamic defects [51–53]. Thermal cycling

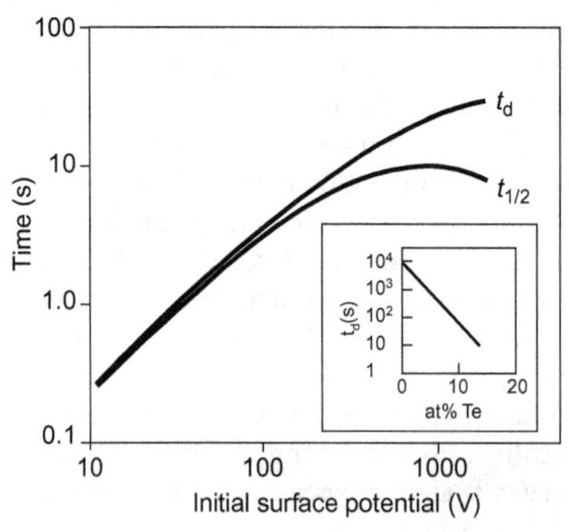

Figure 6.1 Log–log plots of the depletion time t_d and time for the surface potential to decay to its half value $t_{1/2}$ versus charging voltage V_0 for an amorphous Se: Te 13 wt% photoreceptor film of thickness 70 μm. The inset shows the dependence of the depletion time t_d on the Te content. *Source*: From Ref. [48].

experiments show that the response of the depletion time to temperature steps is retarded, as would be expected when the structure relaxes toward its metastable, liquid-like equilibrium state. The only possible inference is that t_d must be controlled by structure-related thermodynamic defects. The generation of such defects is therefore thermally activated. We should note that since the depletion-discharge mechanism involves the thermal emission of carriers from deep localized states, it is strongly temperature dependent. Therefore, the location of these states in the mobility gap can be easily determined. For example, t_d increases in an approximate Arhcnian fashion with decreasing temperature. In addition (to the deterioration of the dark decay), there is an increase in the residual potential for a-$Se_{1-x}Te_x$ alloys.

6.5 Residual Potential

Figure 6.2 displays the $\mu\tau$ product for holes and electrons. This parameter was determined from xerographic residual potential in a-$Se_{1-x}Te_x$ monolayer films. Even with very little Te alloying there is a considerable rise in both hole and electron deep traps. The relationship between $\mu\tau$ product and the residual potential has been evaluated by numerous authors (see, e.g., [5] and references cited). When the Te concentration exceeds 12 wt% Te, the residual potential is more than an order of magnitude larger than typical values for pure selenium.

There are two key parameters which are of interest. The first residual, V_{R1}, is related to the $\mu\tau$ product via [2]$V_{R1} = L^2/2\mu\tau$, which assumes strong injection and weak trapping.

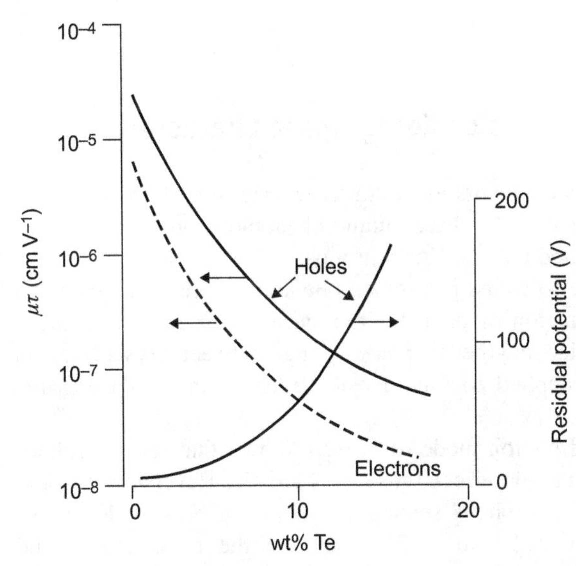

Figure 6.2 Hole and electron drift mobility-lifetime product $\mu\tau$ and residual potential versus Te content in a-$Se_{1-x}Te_x$ films. The $\mu\tau$ product was xerographically measured by Abkowitz and Markovics [46].

The repetition of xerographic cycle leads to the eventual saturation of the residual potential. The saturated value corresponds to the entire deep trap in the bulk being filled and occurs typically after 100 xerographic cycles. The trap-filling interpretation can be readily verified by checking whether $V_{R\infty}$ shows any dependence on cycling frequency to exclude a dynamic equilibrium condition involving trap filling and emptying. The saturated residual potential is simply given by:

$$V_{R\infty} = eN_tL^2/2\varepsilon_0\varepsilon \tag{6.34}$$

where N_t is the deep-trap concentration. The results from the saturated residual potential can be used in conjunction with IFTOF lifetime values to evaluate the capture radius of traps. Application of ballistic and diffusional trapping models of Street [54] to the IFTOF lifetime and cycled-up residual-potential data imply capture radii of 2–3 Å for both pure a-Se and a-As_xSe_{1-x} alloys. The latter do not necessarily have to be paramagnetic neutral centers since an IVAP would effectively look neutral to a drifting hole [19].

Both the first residual and the cycle-up saturated residual potential, V_{R1} and V_{RS}, are sensitive to preillumination as well as to temperature and alloying. For example, when a-Se films are preilluminated with white light, the buildup of the residual potential occurs more rapidly toward a much higher saturated residual potential. Furthermore, the parameters mentioned (residual potentials V_{R1} and V_{RS}) increase with exposure time. The effects of preillumination with band-gap light on the main xerographic characteristics for the case of pure amorphous selenium were examined in details by Abkowitz and Enck in their extensive work [32] and, for Se-rich a-As_xSe_{1-x} and a-Sb_xSe_{1-x} alloys, by the present authors [49,50,55–65]. Clearly, exposure to white light generates an appreciable concentration of deep hole traps. The effects of preillumination xerographic parameters we consider in the following section.

6.6 Photoinduced Changes of Xerographic Characteristics

Amorphous chalcogenide semiconductors exhibit a large spectrum of photoinduced phenomena. According to Tanaka [66], these unique phenomena observed in chalcogenide glasses can be classified into two main groups.

The first group includes the so-called heat-mode phenomena: the heat generated through nonradiative recombination of photoexcited carriers caused atomic structural changes. The most familiar may be the phase change between crystalline and amorphous phases, which is applied to high-density (more than 1 GB) erasable optical memories.

The second is the so-called photon mode. It is well known that for amorphous chalcogenide semiconductors unique effects are characteristic. Reversible thermal and photoinduced changes in amorphous semiconductors have been under active investigation during several decades [66–72]. They are the most unique and intriguing features of chalcogenide vitreous semiconductors. Photoinduced changes appear as nearly parallel shift of the optical absorption edge to lower energy or a reduction in band-gap on exposure to light (so-called "photodarkening"). Such irradiation also causes a change in various physical properties. The physical origin of the processes which takes place at reversible transformations induced by band-gap light and thermal annealing remains unclear. Particular interest is aroused when observations suggest that these changes affect photoelectronic behavior. It is important to note that analysis of changes in photoelectronic properties can be used to relate photoinduced effects to specific changes in the electronic density of gap states.

During the last decades there has been, however, very limited systematic study of photoinduced effects on states, especially of deep lying states in the band gap. In addition, most photodarkening studies have focused upon prototypical binary (and ternary) chalcogenide alloys and relatively little are known about the characteristics of elemental and chalcogen-rich glasses. At the same time, elemental and chalcogen-rich amorphous semiconductors serve as useful model systems for studying the influence of photodarkening on physical properties. Therefore, we consider in the following photoinduced changes of deep levels in pure and selenium-rich amorphous layers.

Careful analysis of the time- and temperature-dependent decay of the surface voltage on an amorphous film can be used to map the density of states. This procedure can be performed after corona charging of the sample but prior to exposure (known as xerographic dark decay) or after exposure (PID), in the first cycle or after many cycles. The procedure is illustrated for a-As_xSe_{1-x} where residual potential can be measured without complication. Amorphous As_xSe_{1-x} ($0<x<0.20$) is found to be characterized by relatively discrete gap state structure. Measurements performed by the present authors readily discern photo- and thermostructural effects on gap state population (see, e.g., [49,50,55–65] and references therein). Thus, during these structural transformations systematic variation in the density of localized states distributed throughout the mobility gap are observed. This observation is consistent with the view that native defects play a key role in photoelectronic behavior of amorphous chalcogenides.

The illumination of amorphous As_xSe_{1-x} films by light with energy near the optical gap causes changes in basic electrophotographic parameters: dark discharge rate, initial charging potential, residual potential, and its dark decay rate. The initial charging voltage V_0 becomes smaller and the surface potential decay rate dV/dt increases after previous photoexcitation. It must be emphasized that, depending on composition, ordinary (changes of either V_0 or dV/dt) or complex (simultaneous changes of V_0 and dV/dt) photoinduced effects may be observable. The former take place in the range of 2–6 at% As, whereas for the latter concentrations above 8 at % As are needed.

In pure selenium the photoinduced change of is comparatively small. Initially parameter γ increases with the arsenic addition, then, for $x>0.20$, decreases. This decrease can be caused by increasing dV/dt with As content for dark-rested samples. It is of interest to point out that the "memory" effect (the time interval during which the changes in preexcited film parameters may be observable) appreciably prolonged with As content: from ≈ 40 min in pure selenium to ≈ 10 h in $As_{0.2}Se_{0.8}$. The ratio of discharge rates in preilluminated and dark-rested samples, $\gamma = (dV/dt^*)/(dV/dt)$, is essentially composition dependent.

It is convenient to represent the persistence of light-induced deep trapping by a relaxation function. The latter is determined as follows: the specimen is subjected to a fixed exposure, and then allowed to dark rest for a time τ which is experimentally varied. The relaxation functions $\varphi(\tau)$ and $\zeta(\tau)$ determined as shown below. An increasing the time interval of dark adaptation (i.e., the dark-resting time of an exposed film before charging it to a certain potential and recording the decay of surface potential) causes a diminishing of the observable photoinduced changes on xerographic parameters. By analogy to the photocurrent transients [55,56,58], the relaxation functions, $\varphi(\tau) = [V_0 - V_0^{\tau}]/[V_0 - V_0^*]$ and $\zeta(\tau) = [(dV^*/dt) - dV^{\tau}/dt]/[(dV^*/dt - dV/dt]$, were estimated. The two relaxation functions characterize the recovery of initial charging potential and the dark decay rate with the dark-resting time after irradiation is terminated.

It is of particular significance that "memory" effects are influenced not only by variation of composition but also by electric fields [56,58,63–65]. Clearly, the photoinduced change of the dark discharge rate may successfully be "frozen-in" by applying an electric field $E \approx 3 \times 10^5$ V/cm immediately after light exposure.

Dark decay curves of the surface potential in double-logarithmic representation display two distinct rate processes $dV/dt \sim t^{-(1-p)}$ and $dV/dt \sim t^{-(1+p)}$ with crossover from one regime to the other at $t = t_d$. Detailed analysis of the discharge process shows that t_d shifted to smaller values with increasing As content and after exposure. Note that under low charging voltages the surface potential at time t_d was approximately half the initial value V_0. The features observed in the dark decay of As_xSe_{1-x} alloy films can be completely accounted for by a depletion-discharge model [31,37–44]. So, the bulk process driving dark decay is emission and sweep-out of holes from states near mid-gap leading to progressive formation of negative space–charge. The rise in dark discharge rate and the shift of t_d with arsenic concentration and photoexcitation may be due to enhanced thermal generation of holes from deep centers. From the temperature dependence of t_d, it is estimated that the emitting sites are located at 0.8–0.9 eV above the valence band mobility edge.

6.7 Fatigue Effects in Se-Rich Photoreceptors

The repetition of the xerographic cycle leads to the saturation of the residual potential. For the films under examination, the residual voltage V_R increases with As content. For example, addition of 2 at% As to a-Se leads to a change in the first-cycle residual voltage from 1.9 to 5.0 V which is equivalent to a change of the carrier range from 2.6 to 1.1×10^{-7} cm^2 V^{-1}. Substituting $\mu_e(\text{Se}) = 4.9 \times 10^{-3}$ cm^2 V^{-1} s^{-1} and $\mu_e(As_{0.02}Se_{0.98}) = 6.0 \times 10^{-4}$ cm^2 V^{-1} s^{-1} into $V_{R1}/V_0 = L^2/(2\mu\tau V_0)$, we find carrier lifetimes $\tau \approx 5.3 \times 10^{-5}$ s^{-1} and $\tau \approx 1.8 \times 10^{-4}$ s^{-1} in a-Se and $As_{0.02}Se_{0.98}$, respectively.

Preillumination also leads to V_{R1} increasing in Se up to 3.7 V. This is caused by a diminishing of $\mu\tau$ to $\approx 1.3 \times 10^{-7}$ cm^2 V^{-1}. Taking into account the invariance of $\mu_e = 4.9 \times 10^{-3}$ cm^2 V^{-1} s^{-1} with light exposure, it is obvious that the lifetime reduction (2.6×10^{-5} s) is the only reason for photoinduced change.

An increase in V_{RS} in preexposed films indicates photoenhanced accumulation of charge at deep centers. We obtain $N_t \approx 10^{15}$ cm^{-3} and $N_t^* \approx 4 \times 10^{15}$ cm^{-3} for dark-rested and preexposed $As_{0.1}Se_{0.9}$ films, respectively.

The residual potential decays to zero. This process is controlled by the spectrum of trap release times. The trap energies can be deduced from an analysis of isothermal residual potential decay curves using

$$V_{RS} = \sum C_i \exp(-t/\tau_i) \tag{6.35}$$

where $\tau_i = \nu_i \exp(-E_i/kT)$ is the release time from the ith trap, ν_i is the frequency factor, and E_i is the trap depth. We find that deep levels in amorphous selenium reside at $E_i = 0.85$ eV and $E_i = 1.0$ eV for holes and electrons, respectively. Their depth becomes somewhat shallower with addition of As, for example, $E_i = 0.80$ eV and $E_i = 0.90$ eV for holes and electrons in $As_{0.1}Se_{0.9}$. The more rapid decay of the residual voltage in preilluminated films relative to dark-rested film indicates a slight decrease in the depth of those states. A comparison of room-temperature recovery in TOF and xerographic measurements demonstrates that relaxation of deep centers in preilluminated (exposed) films occurs on the same time scale. In other words, the deep gap centers which control the xerographic dark decay and residual voltage are, like the trapping centers discussed in TOF experiments, characteristically metastable.

Band-gap light can, in principle, have two distinct effects on the electronic structure of the mobility gap. Band-gap light can either introduce (generate) new localized states or initiate conversion of traps of small cross section to traps of larger cross section [29,31]. Consequently, the latter become accessible in deep level spectroscopy only after irradiation. In that sense we may also consider such converted localized states as "new" localized states (created by irradiation).

Principal results are summarized as follows:

1. The xerographic process is a sensitive and relatively simple technique for studying deep gap states in wide-gap amorphous semiconductors.
2. There is important information yielded at each stage of the xerographic cycle. The bulk process controlling dark decay is demonstrated to be thermal emission of holes from states deep in the mobility gap. The hole sweep-out process causes the progressive development of negative deeply trapped space–charge in the bulk. The process is experimentally manifested in the distinct shape of the associated dark discharge curves. The rate of dark discharge and the detailed shape of the dark decay vary systematically with composition. Phenomenological parameters of the dark decay to emission center distribution in the gap.
3. The analysis and interpretation of the dark decay, first-cycle residual potential and cycled-up saturated residual potential are considered. It is shown that the charge-carrier lifetime, τ, the range of the carriers, $\mu\tau$, and the integrated concentration of deep traps in

the mobility gap can be readily and accurately determined. Xerographic measurements on Se-rich amorphous photoconductors have indicated the presence of relatively narrow distribution of deep hole traps with integrated density of about 10^{13} cm^{-3}. These states are located at ~0.85 eV from the valence band. A good correlation was observed between residual potential and the hole range, in agreement with the simple Warter expression.

4. The capture radius is estimated to be $r_c = 2 - 3$ Å. Since r_c for pure a-Se and a-As_xSe_{1-x} is comparable to the Se–Se interatomic bond length in a-Se, it can be suggested that deep hole trapping centers in these chalcogenide semiconductors are neutral-looking defects, possibly of IVAP in nature.

5. Both the injected-hole and injected-electron range are reduced (i.e., deep trapping is increased) in amorphous selenium-based films which have undergone prior photoexcitation with near-band-gap light. This photoenhanced trapping is metastable, decaying away as the film dark rests. Recovery process becomes slower with increasing As content. Qualitative explanation of the observed behavior may be based on associating the deep states with $C_3^+$and C_1^- IVAP centers.

References

[1] S.O. Kasap, J.A. Rowlands, in: P. Boolchand (Ed.), Insulating and Semiconducting Glasses, World Scientific Publishing Co. Ltd., Singapore, 2000, pp. 782–808.
[2] P.J. Warter, Appl. Optic. Suppl. 3 (1969) 65.
[3] K.K. Kanazawa, I.P. Batra, J. Appl. Phys. 43 (1972) 1845.
[4] S.O. Kasap, V. Aiyah, B. Polischuk, M. Abkowitz, Philos. Mag. Lett. 62 (1990) 377.
[5] S.O. Kasap, V. Aiyah, B. Polischuk, A. Bhatacharyya, Z. Liang, Phys. Rev. B 43 (1991) 6691.
[6] W.E. Spear, J. Non-Cryst. Solids 1 (1969) 197.
[7] J.M. Marshall, Rep. Prog. Phys. 46 (1983) 1235.
[8] S.O. Kasap, B. Polischuk, D. Doods, Rev. Sci. Instrum. 61 (1990) 2080.
[9] J. Mort, A. Troup, M. Morgan, S. Grammatica, J.C. Knights, R. Lujan, Appl. Phys. Lett. 38 (1981) 277.
[10] S.O. Kasap, C. Juhasz, Solid State Commun. 63 (1987) 553.
[11] F.K. Dolezalek, W.E. Spear, J. Phys. Chem. Solids 36 (1975) 819.
[12] Z. Hecht, Z. Phys. 77 (1932) 235.
[13] W.E. Spear, J. Mort, Proc. Soc. 81 (1963) 130.
[14] R.A. Street, Appl. Phys. Lett. 41 (1982) 1060.
[15] R.A. Street, J. Zesch, M.I. Thompson, Appl. Phys. Lett. 43 (1983) 1425.
[16] R.A. Street, J. Zesch, M.I. Thompson, Appl. Phys. Lett. 43 (1983) 672.
[17] R.A. Street, C.C. Tasi, M. Stutzman, J. Kakalios, Philos. Mag. B 56 (1987) 289.
[18] W.E. Spear, W.E. Steemers, H. Mannsperger, Philos. Mag. B 48 (1983) L49.
[19] B.T. Kolomiets, E.A. Lebedev, Sov. Phys. Solid State 8 (1966) 905.
[20] I.C. Schottmiller, M. Tabak, G. Lucovsky, A. Ward, J. Non-Cryst. Solids 4 (1970) 80.
[21] M.D. Tabak, W.J. Hillegas, J. Vac. Sci. Technol. 9 (1972) 387.
[22] D.M. Pai, in: J. Stuke, W. Brenig (Eds.), Amorphous and Liquid Semiconductors. Proc. 5th Intern. Conf. on Amorphous and Liquid Semiconductors (Garmish-Partenkirchen, Taylor & Francis, London, 1974, p. 355.
[23] J.M. Marshall, F.D. Fisher, E.A. Owen, Phys. Stat. Solidi 25 (1974) 419.
[24] T. Takahashi, J. Non-Cryst. Solids 34 (1979) 307.

[25] S.D. Baranovsky, E.A. Lebedev, Sov. Phys. Semicond. 19 (1985) 635.
[26] E.A. Owen, W.E. Spear, Phys. Chem. Glasses 17 (1974) 174.
[27] S.M. Vaezi-Nejad, C. Juhasz, J. Int, Electronics 67 (1989) 437.
[28] R.M. Schaffert, Electrophotography. Society of Photographic Scientists and Engineers, Focal Press, London, 1975.
[29] S.M. Vaezi-Nejad, Int. J. Electronics 62 (1987) 361.
[30] L.B. Schein, Electrophotography and Development Physics, Springer-Verlag, New York, NY, 1988.
[31] M. Abkowitz, R.C. Enck, Phys. Rev B 25 (1982) 2567.
[32] M. Abkowitz, R.C. Enck, Phys. Rev. B 27 (1983) 7402.
[33] M.D. Tabak, P.J. Warter, Phys. Rev. 173 (1968) 899.
[34] F.K. Dolezalek, in: J. Mort, D.M. Pai (Eds.), Photoconductivity and Related Phenomena, Elsevier, New York, NY, 1976, p. 27.
[35] S.W. Ing, J.H. Neyhart, J. Appl. Phys. 43 (1972) 2670.
[36] L.B. Schein, Phys. Rev. B 10 (1974) 3451.
[37] E. Montrimas, S. Tauraitiene, A. Tauraitis, in: W.F. Berg, K. Klauff (Eds.), Current Problems in Electrophotography, Walter de Cruyter, New York, NY, 1972, p. 138.
[38] M. Abkowitz, S. Maitra, J. Appl. Phys. 61 (1987) 1038.
[39] A.R. Melnyk, J. Non-Cryst. Solids 35–36 (1980) 837.
[40] M. Abkowitz, G.M.T. Foley, J.M. Markovics, A.C. Palumbo, Appl. Phys. Lett. 46 (1985) 393.
[41] M. Baxendale, C. Juhasz, SPIE Proc. 1253 (1990).
[42] S.O. Kasap, M. Baxendale, C. Juhasz, IEEE Trans. Indust. Appl. 27 (1991) 620.
[43] S.O. Kasap, J. Electrostat. 22 (1989) 69.
[44] M. Abkowitz, F. Jansen, A.R. Melnyk, Philos. Mag. B 51 (1985) 405.
[45] M. Abkowitz, J. Non-Cryst. Solids 66 (1984) 315.
[46] M. Abkowitz, J.M. Markovics, Solid State Commun. 44 (1982) 1431.
[47] H. Scher, E.W. Montroll, Phys. Rev. B 12 (1974) 2455.
[48] S.O. Kasap, in: A.S. Diamond, D.S. Weiss (Eds.), Handbook of Imaging Materials, Second Addition, Marcel Dekker, Inc., New York, NY, 2002.
[49] V.I. Mikla, I.P. Mikhalko, Y.U. Nagy, A.V. Mateleshko, J. Mater. Sci. 35 (2001) 4907.
[50] V.I. Mikla, I.P. Mikhalko, Y.U. Nagy, Mater. Sci. Eng. B64 (1999) 1.
[51] M. Abkowitz, Polym. Eng. Sci. 24 (1984) 1140.
[52] M. Abkowitz, J. Non-Cryst. Solids 97–98 (1987) 1163.
[53] M. Abkowitz, D. Adler, H. Fritzsche, S.R. Ovshinsky (Eds.), Physics of Disordered Materials, Plenum Press, New York, 1984.
[54] R.A. Street, Philos. Mag. B 49 (1984) L15.
[55] V.I. Mikla, D.G. Semak, A.V. Mateleshko, A.R. Levkulich, Sov. Phys. Semicond. 21 (1987) 266.
[56] V.I. Mikla, D.G. Semak, A.V. Mateleshko, A.A. Baganich, Phys. Status Solidi A 117 (1990) 241.
[57] V.I. Mikla, D.G. Semak, A.V. Mateleshko, A.R. Levkulich, Sov. Phys. Semicond. 23 (1989) 80.
[58] V.I. Mikla, J. Phys. Condens. Matter. 8 (1996) 429.
[59] V.I. Mikla, I.P. Mikhalko, Y.U. Nagy, J. Non-Cryst. Solids 142 (2001) 1358.
[60] V.I. Mikla et al., 14th General Conf. GCMD-14, Madrid, 1994, p. 1055.
[61] V.I. Mikla, et al., 15th General Conf. of the Condensed Matter Devision, Baveno-Stresa, Lago Maggiore, Italy, 1996, 20, p. 103.

[62] V.I. Mikla, Int. Workshop on Advance Technology of Multicomponent Solid Films, Uzhgorod, 1996, Book of Abstracts, p. 84.
[63] V.I. Mikla, et al., Patent USSR 4088420, 1986.
[64] V.I. Mikla, et al., Patent USSR 4273793, 1987.
[65] V.I. Mikla, et al., Patent USSR 4638891, 1988.
[66] K. Tanaka, Curr. Opin. Solid State Mater. Sci. 1 (1996) 567.
[67] V.M. Lyubin, in: D. Adler, M. Kastner, H. Fritzsche (Eds.), Physics of Disordered Materials, Plenum Press, New York, NY, 1985, p. 673.
[68] K. Tanaka, Phys. Status Solidi B 246 (2009) 1744.
[69] A.V. Kolobov, Photo-induced Metastability in Amorphous Semiconductors, Willey-VCH, Weinheim, 2003.
[70] V.I. Mikla, V.V. Mikla, Metastable States in Amorphous Chalcogenide Semiconductors, Springer, Heidelberg, 2009.
[71] K. Tanaka, C. Glasses, Encyclopedia of Materials: Science and Technology, Elsevier Science Ltd, United Kingdom, 2001, pp. 1123–1131.
[72] V.I. Mikla, V.V. Mikla, Trap Level Spectroscopy in Amorphous Chalcogenide Semiconductors, Elsevier Insights, New York, NY, 2010.

7 Ultrasound Imaging

7.1 Introduction

For a considerable number of years after Roentgen's discovery of X-rays, the use of ionizing radiation for diagnostic imaging remained the only method for visualizing the interior of the body. Fortunately, during the second half of the twentieth century new imaging methods, including some based on principles totally different from those of X-rays, were discovered. Ultrasonography was one such method that showed particular potential and greater benefit than X-ray-based imaging. During the last decade of the twentieth century, use of ultrasonography became increasingly common in medical practice and hospitals around the world. The benefit and even the superiority of ultrasonography over commonly used X-ray techniques, resulting in significant changes in diagnostic imaging procedures.

7.2 History—Milestones

Modern acoustics as we know it today was put forth in the classic 1877 work The Theory of Sound by Lord Rayleigh [1].

In 1880, French physicists Pierre and Jacques Curie discovered the piezoelectric effect [2]. French physicist Paul Langevin attempted to develop piezoelectric materials as senders and receivers of high-frequency mechanical disturbances (ultrasound waves) through materials. Langevin developed one of the first uses of ultrasound for underwater echo ranging of submerged objects with a quartz crystal at an approximate frequency of 150 kHz [3]. He was, perhaps, the first to observe that ultrasonic energy could have a detrimental effect upon biological material.

His specific application was the use of ultrasound to detect submarines during World War I. This technique, sound navigation and ranging (SONAR), finally became practical during World War II. Industrial uses of ultrasound began in 1928 with the suggestion of Soviet Physicist Sokolov that it could be used to detect hidden flaws in materials. The reader may find detailed history in excellent review by W. O'Brian [4].

Medical uses of ultrasound through the 1930s were confined to therapeutic applications such as cancer treatments and physical therapy for various ailments. Diagnostic applications of ultrasound began in the late 1940s through collaboration between physicians and engineers familiar with SONAR [5–8].

Medical Imaging Technology. DOI: http://dx.doi.org/10.1016/B978-0-12-417021-6.00007-1

Another decade passed before a more detailed, experimental study was conducted by Wood and Loomis to investigate Langevin's 1917 observation [9]. Although the ultrasonic levels were not specified, their experimental studies showed that ultrasonic energy had a range of effects from rupture of *Spirogyra* and *Paramecium* to death of small fishes and frogs by a one- to two-minute exposure, the latter also observedå by Langevin with a Poulsen arc oscillator. Considerable work followed and in the earliest review paper on this subject, Harvey reported on the physical, chemical, and biological effects of ultrasound in which alterations were produced in macromolecules, microorganisms, cells, isolated cells, bacteria, tissues, and organs with a view toward the identification of the interaction mechanisms [10]. The ultrasonic exposure conditions of these early works were neither well characterized nor reported, but the exposure levels were undoubtedly high.

It is not known when scientists initially recognized the two principal biophysical mechanisms that are currently invoked, i.e., thermal and cavitation. The application of ultrasound to therapeutically heat tissue was suggested as early as 1932 [11]. Also in the 1920s, Boyle and his colleagues were perhaps the first to observe ultrasound-produced gas bubble formation in liquids and recognized this phenomenon as cavitation [12–14]. The ultrasound-induced biological observations by Langevin and by Wood and Loomis were caused by gas bubbles as demonstrated with overpressurization experiments [15]. Ultrasound-induced tissue heating was applied extensively as a therapeutic agent in the 1930s and 1940s [16]. However, while it was clear that ultrasound could effectively heat tissue, and excess enthusiasm resulted in numerous clinical applications being proposed and tried, the inferior clinical experience caused this modality to fall into disfavor. Thus, as Hill [17] observed of this time, "it is perhaps unfortunate that the generation of ultrasound proved to be so relatively simple and cheap that a considerable practice was built up." During this time period, with an understanding that ultrasound at sufficient levels could have a dramatic effect on tissues, and produce large temperature increases, the potential for ultrasonic surgery was proposed. This ability to noninvasively burn focal tissue volumes deep in the body using ultrasound was first proposed in 1942 [18,19] as a neurosurgery technique. Ultrasound surgery and its biophysical mechanism (heating) were further developed in the late 1940s and early 1950s [20]. Also proposed in 1948 and applied in 1952 was the application of ultrasound surgery to destroy the vestibular function to treat the symptoms of Menière's disease [21].

There was very little activity for about a decade in the development of ultrasound imaging capabilities following Langevin's SONAR work during the World War I. In 1928, Sokolov proposed and a few years later demonstrated a through-transmission technique for flaw detection in metals [22]. Firestone's 1942 patent [23] for flaw detection in metals, and his later demonstration, is considered the first modern pulse-echo ultrasound technique for flaw detection, [24,25] and the basis for pulse-echo imaging in medicine. A technique that was never developed into clinical application was based on the principle of differential attenuation, a through-transmission ultrasonic technique which was constructed in 1937 for brain imaging by the Dussik brothers (Figures 7.1 and 7.2), [26] and by others in the late 1940s [27].

Figure 7.1 Paul Langevin (1872–1946).

Figure 7.2 Karl Theo (Theodore) Dussik 1908–1968.

The development of diagnostic ultrasound instrumentation as we know it today was initiated around the time of the end of the World War II, a time when fast electronic circuitry was becoming available as a result of the wartime RADAR and SONAR efforts, both of which utilized the pulse-echo principle. In the late 1940s and early 1950s, Howry showed that tissue interfaces could be detected in ultrasound echoes, Wild [28–31] showed that tissue structure could be differentiated (cancer from benign) in ultrasound echoes, and Ludwig and Struthers [32] showed that gall stones could be detected in ultrasound echoes, these being A-mode applications; and Howry and Bliss [33] and Wild and Reid [34,35] independently built and successfully demonstrated the earliest B-mode, bistable, ultrasound scanners.

Transduction materials had a central role in the diagnostic ultrasound developments. Quartz-based transducers had limited application. Quartz was used because other transduction devices and/or materials had not been developed, and it possessed high mechanical strength and low internal friction. However, large amplitude voltages were required to drive quartz transducers. Around the 1930s, quartz started to be replaced for underwater ultrasound applications by Rochelle salt and was used up to the time of the World War II [36] at which time it was replaced with piezoelectric ceramics, the first of which was barium titanate [37]. The strongly piezoelectric Rochelle salt was replaced because its properties were very susceptible to moisture.

Piezoelectric ceramics had two major advantages, improved efficiency to convert electrical into acoustic energy and could be processed into varying shapes and sizes. Also, piezoelectric ceramics did not require large amplitude voltages to drive them. Pure barium titanate's principal disadvantage was its low-temperature Curie point (−5°C) and to improve this, small amount of calcium titanate and/or lead titanate were added. A major advancement in transducer materials occurred in 1954 when Jaffe et al. [38] discovered lead−titanate−zirconate compositions, a piezoelectric material which retained most of the advantageous properties of the titanates while extending its operating temperature range.

Interestingly, both Howry's first system and Wild's system in the late 1940s used quartz as the transduction material. However, Howry's later systems employed some of the newer transduction materials [33] and Reid and Wild [39] had evaluated the applicability of an annular array made from barium titanate.

The scientific and engineering advances during the 1940s had enormous influences on the development of diagnostic ultrasound. In 1943, the Moore School of Electrical Engineering at the University of Pennsylvania was contracted to construct the Electronic Numerical Integrator and Computer (ENIAC), the first electronic computer with design specifications to calculate 5000 operations per second. Up to then, the only computing capabilities were basically analog computers. Another major program was the radar activity at the Radiation Laboratory at MIT from which the need was identified for fast electronic circuitry in order to achieve higher frequencies for better imaging resolution.

Radar was also responsible for much of the diagnostic ultrasound nomenclature. In the mid-1930s, the American Telephone and Telegraph Company identified a need to develop fast electronic switching to replace electromechanical switching. This lead AT&T's Bell Labs to initiate a solid-state research program which lead to the invention of the point-contact transistor for which its fiftieth anniversary was celebrated in December, 1997.

John Julian Wild, a Cambridge medical graduate, laid the foundations of ultrasonic tissue diagnosis with the publication of A-mode (amplitude mode) ultrasound investigations of surgical specimens of intestinal and breast malignancies, the development of a linear handheld B-mode (brightness mode) instrument and early descriptions of endoscopic (transrectal and transvaginal) A-mode scanning transducers in 1955 [37,38].

A key figure in the development of medical ultrasound in clinical practice was Professor Ian Donald of Glasgow. Having gained initial experience in radar and sonar techniques while serving in the Royal Air Force during World War II, he was enthused in medical ultrasound on meeting John Wild while he was working at the Hammersmith in London. On becoming the Regis Professor of Midwifery of the University of Glasgow, Ian Donald and coworkers began a series of studies that would establish a role for medical ultrasound, overcoming initial clinical skepticism from his colleagues who believed that manual abdominal and pelvic examination provided sufficient diagnostic certainty. With the help of the engineering firm Kelvin Hughes Ltd, Ian Donald used a "flaw detector" to differentiate cystic and solid abdominal masses—in one case altering a clinical diagnosis of terminal carcinoma to simple ovarian cyst—leading to the publication of their findings in the

Lancet in 1958, a major milestone in medical ultrasound [40–42]. With his colleagues, Donald first developed a two-dimensional scanner and then an automatic scanner in 1960, made the first ante-partum diagnosis of placenta previa using ultrasound, developed the method for measuring the biparietal diameter of the fetal head in 1962 and was the first to utilize the full bladder to allow the detection of very early pregnancy of about 6–7 weeks gestation in 1963 [40–42].

In the years that follow essential developments were made:

- The first grayscales images were produced in 1950.
- In real time by Siemens device in 1965.
- Electronic beam-steering using phased-array technology in 1968.
- Popular technique since mid-1970s.
- Substantial enhancements since mid-1990.

7.3 Basic Physics

7.3.1 Definition

Ultrasound is the term used to describe sound of frequencies above 20,000 Hertz (Hz), beyond the range of human hearing. Frequencies of 1–30 megahertz (MHz) are typical for diagnostic ultrasound (Figure 7.3).

Diagnostic ultrasound imaging depends on the computerized analysis of reflected ultrasound waves, which noninvasively build up fine images of internal body structures. The resolution attainable is higher with shorter wavelengths, with the wavelength being inversely proportional to the frequency. However, the use of high frequencies is limited by their greater attenuation (loss of signal strength) in tissue and thus shorter depth of penetration. For this reason, different ranges of frequency are used for examination of different parts of the body (Figure 7.3):

- 3–5 MHz—for abdominal areas
- 5–10 MHz—for small and superficial parts and
- 10–30 MHz—for the skin and the eyes.

7.3.2 Generation of Ultrasound

Piezoelectric crystals or materials are able to convert mechanical pressure (which causes alterations in their thickness) into electrical voltage on their surface (the piezoelectric effect). Conversely, voltage applied to the opposite sides of a

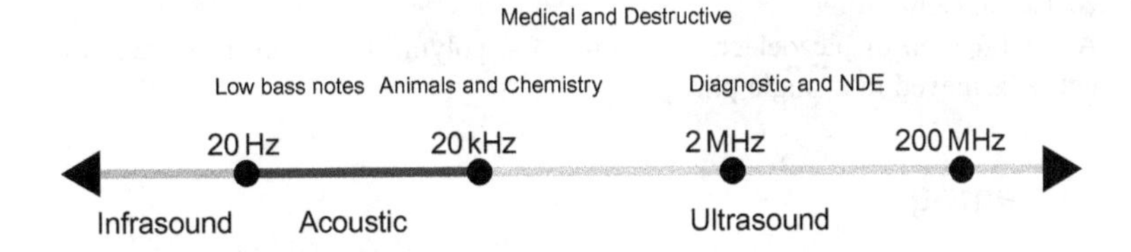

Figure 7.3 Ultrasound range diagram.

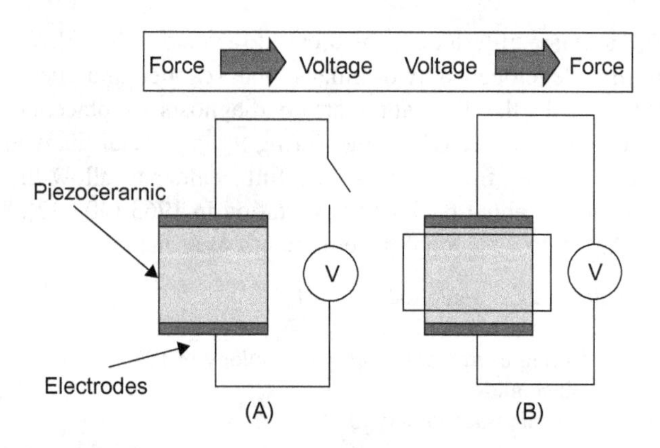

Figure 7.4 Piezoelectricity means "pressure electricity," which is used to describe the coupling between a material's mechanical and electrical behaviors. Piezoelectric effect—when a piezoelectric material is squeezed or stretched, electric charge is generated on its surface. Conversely, when subjected to electric voltage input, a piezoelectric material mechanically deforms. "Piezo" is Greek for pressure. Piezoelectricity refers to the generation of an electrical response to applied pressure. Pierre Curie used the piezoelectric properties of quartz crystals to construct a device to measure the small changes in mass that accompany radioactive decay.

piezoelectric material causes an alteration in its thickness (the indirect or reciprocal piezoelectric effect). If the applied electric voltage is alternating, it induces oscillations which are transmitted as ultrasound waves into the surrounding medium. The piezoelectric crystal, therefore, serves as a transducer, which converts electrical energy into mechanical energy and vice versa (Figures 7.4 and 7.5).

7.4 Piezoelectric Materials

- Piezoelectric Ceramics (man-made materials)
  - Barium titanate ($BaTiO_3$)
  - Lead–titanate–zirconate ($PbZrTiO_3$) = PZT, most widely used
  - The composition, shape, and dimensions of a piezoelectric ceramic element can be tailored to meet the requirements of a specific purpose
- Piezoelectric Polymers
  - Polyvinylidene flouride (PVDF) film
- Piezoelectric Composites
  - A combination of piezoelectric ceramics and polymers to attain properties which can not be achieved in a single phase.

7.5 Imaging

Ultrasound imaging is based on the "pulse-echo" principle in which a short burst of ultrasound is emitted from a transducer and directed into tissue. Echoes are produced

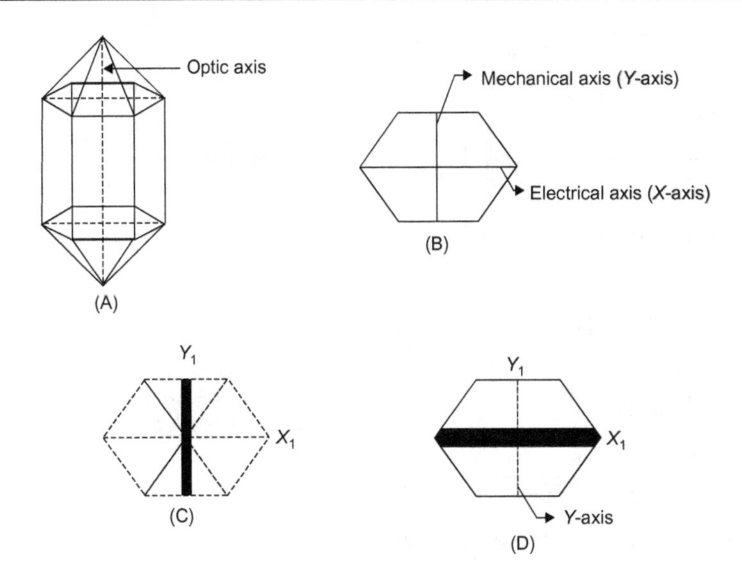

Figure 7.5 Piezoelectric quartz crystals: (A) crystal, (b) electric and mechanical axis, (c) X-cut crystal, and (d) Y-cut crystal.

as a result of the interaction of sound with tissue, and some of these travel back to the transducer. By timing the period elapsed between the emission of the pulse and the reception of the echo, the distance between the transducer and the echo-producing structure can be calculated and an image is formed (Figures 7.6 and 7.7).

In diagnostic imaging, frequencies vary from about 2 MHz for some cardiac, transcranial, and deep abdominal applications, through 10 MHz for the imaging of superficial structures such as blood vessels, to 20 MHz or higher for intravascular imaging. At these frequencies, ultrasound has a wavelength of between 1.5 and 0.08 mm, a dimension which sets a fundamental limit on the potential spatial resolution of the resulting image. Better resolution is associated with a higher ultrasound frequency, but absorption of the sound energy by tissue also increases with frequency.

Optimum imaging is thus obtained by choosing the highest frequency transducer which will permit adequate acoustic penetration to identify the region of interest. To this end considerable effort has been expended to develop technologies which will allow the transducer to be positioned nearer to the structure of interest and hence achieve higher resolution.

Ultrasound transducers are usually made of thin disks of an artificial ceramic material such as PZT. The thickness (usually 0.1–1 mm) determines the ultrasound frequency.

Functions of transducer:

- Used as both transmitter and receiver
- Transmission mode: converts an oscillating voltage into mechanical vibrations, which causes a series of pressure waves into the body
- Receiving mode: converts backscattered pressure waves into electrical signal.

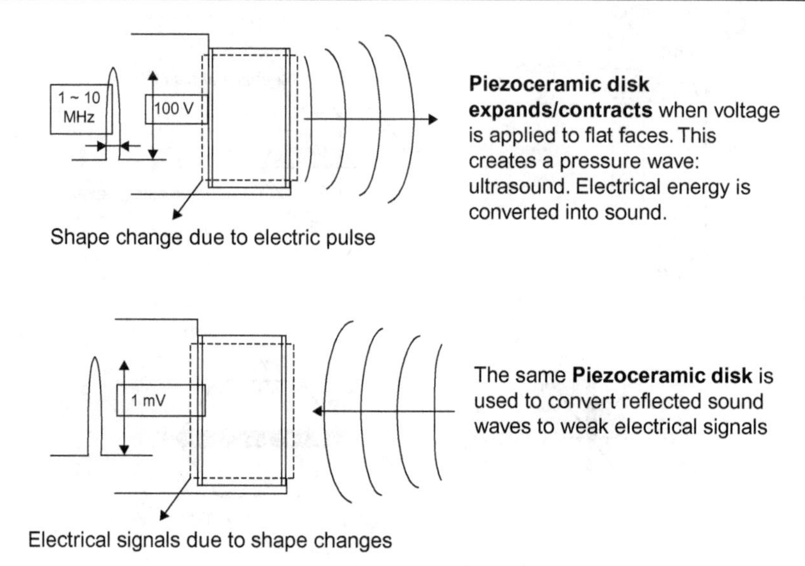

Figure 7.6 Pulse generation (top) and signal detection (bottom).

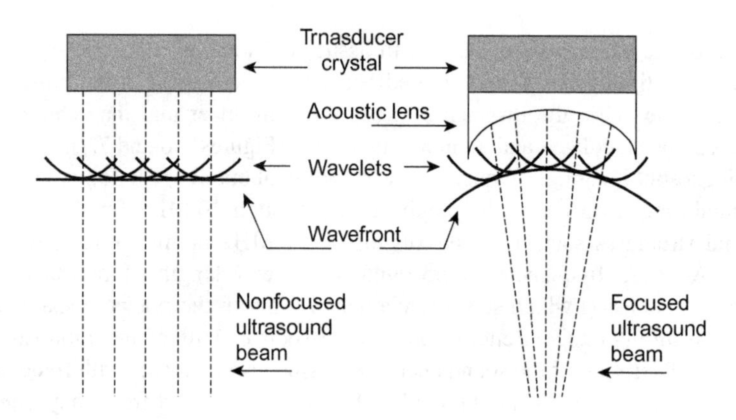

Figure 7.7 Beam forming.

The transducer is the most critical component in any ultrasonic imaging system. In other words, such is the state of the art in systems such as electronic circuitry and display technology that it is the performance of the transducer which determines how closely the limits imposed by the characteristics of the tissues themselves can be approached [43].

Nowadays, the transducers which are in clinical use almost exclusively use a piezoelectric material, of which the artificial ferroelectric ceramic, PZT, is the most common. The ideal transducer for ultrasonic imaging would have a characteristic acoustic impedance perfectly matched to that of the (human) body, have high efficiency as a transmitter and high sensitivity as a receiver, a wide dynamic range and a wide frequency response for pulse operation. PZT has a much higher characteristic impedance than that of water but it can be made to perform quite well by the judicious use of matching layers consisting of materials with intermediate characteristic impedances. Even better performance can be obtained by embedding small particles or shaped structures of PZT in a plastic to form a composite material: this has lower characteristic impedance than that of PZT alone, although it has similar ferroelectric properties.

PVDF is a plastic which can be polarized so that it has piezoelectric properties. The piezoelectric effect can be enhanced by the addition of small quantities of appropriate chemicals. The advantages of this material are that it has a relatively low characteristic impedance and broad frequency bandwidth; it is fairly sensitive as a receiver but rather inefficient as a transmitter.

Piezoelectric transducers are normally operated over a band of frequencies centered at their resonant frequency. The resonant frequency of a transducer occurs when it is half a wavelength in thickness. Typically, a PZT transducer resonant at a frequency of, say, 3 MHz is about 650 μm thick and this means that it is sufficiently mechanically robust for simple, even manual, fabrication techniques to be employed in probe construction. Higher frequency transducers are proportionally thinner and, consequently, more fragile.

The potential of capacitive micromachined ultrasonic transducers (cMUTs) at least partially to replace PZT and PVDF devices in ultrasonic imaging is the subject of current research. A cMUT consists of a micromachined capacitor, typically mounted on a silicon substrate and with a thin electrode membrane as the other plate of the capacitor: these act as the active surface of the transducer. A dc voltage is applied between the plates of the device; the application of an ac voltage causes the membrane to transmit a corresponding oscillatory force, while a received wave causes a corresponding change in the spacing between the plates, thus generating an electrical signal. cMUTs are adequately sensitive as receivers but need high voltages to be effective transmitters. Some of the potential advantages of these devices are that they can be fabricated into arrays with integrated electronics and, if manufactured in large quantities, could be relatively inexpensive.

Although some simple probes contain single-element transducers (e.g., one element for transmitting and one for receiving, in a continuous-wave Doppler system), most modern imaging systems use arrays of transducer elements for beam forming [43].

The basic types of transducers are shown in Figures 7.8–7.10. The block diagram for ultrasound medical imaging system is shown in Figure 7.11.

Arterial diastolic and systolic diameters can be estimated noninvasively using brightness mode (B-mode) or motion mode (M-mode) ultrasound imaging. In real-time B-mode imaging, the amplitude of the reflected wave is reproduced by its brightness on the monitor [44]. Strongly reflecting structures are reproduced as bright points, whereas more weakly reflecting ones are reproduced as dark points. In B-mode, blood reflects very little and the lumen of a blood vessel appears as a

Figure 7.8 Transducers: abdominal obstetrics. The term "transducer" refers to any device that converts energy from one form to another (mechanical to electrical, electrical to heat, etc.).

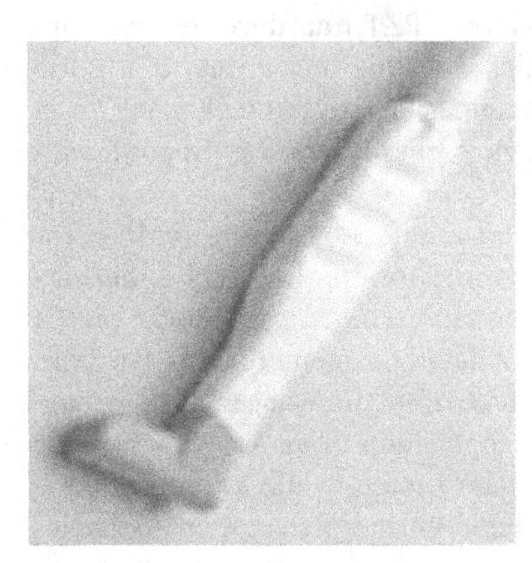

Figure 7.9 Transducers: intraoperative vascular (superficial).

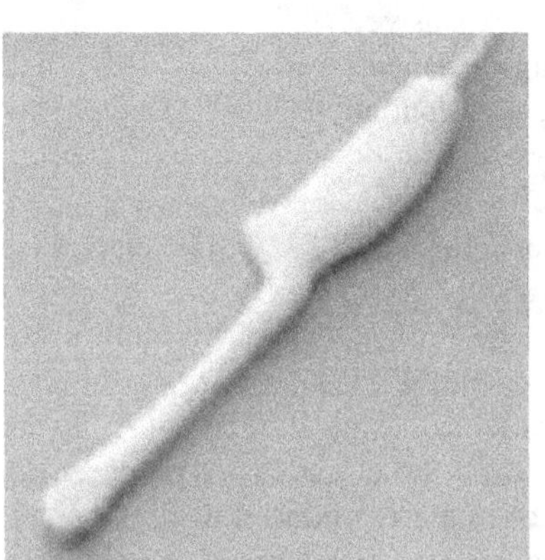

Figure 7.10 Transducers: gynecology obstetrics.

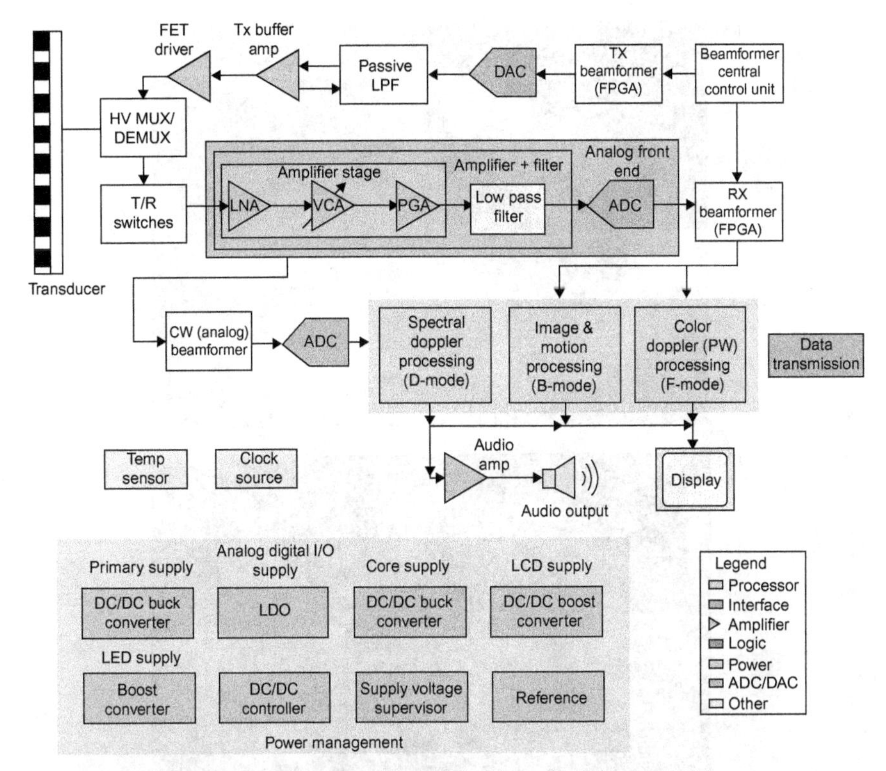

Figure 7.11 Block diagram of medical imaging system.

hypo-echoic band. This enables reliable identification of the investigated vessel. Motion analysis of regions on the arterial lumen from sequences of B-mode images can reveal valuable information about the dispensability of the arterial wall. M-mode can be used to study the movement of interfaces to obtain diagnostic information (Figure 7.12). M-mode involves serial measurements of the location (depth) of a particular echo structure, obtained from periodic pulsing in a single direction, namely along the axis of the transducer.

An M-mode display consists of the time traces of the depth of reflecting interfaces, e.g., arterial walls, over a few cardiac cycles. Maximum and minimum differences between the traces of opposite walls are used as estimates of systolic and diastolic diameters, respectively.

An alternative way to estimate arterial diameters is to automatically segment the vessel lumen and calculate its dimensions. To this end, assumptions should be made about the shape of the lumen. Circular or elliptical shapes have been assumed in previous studies to estimate the arterial distension waveform. In transverse sections of the carotid artery, the arterial lumen can be considered circular, and

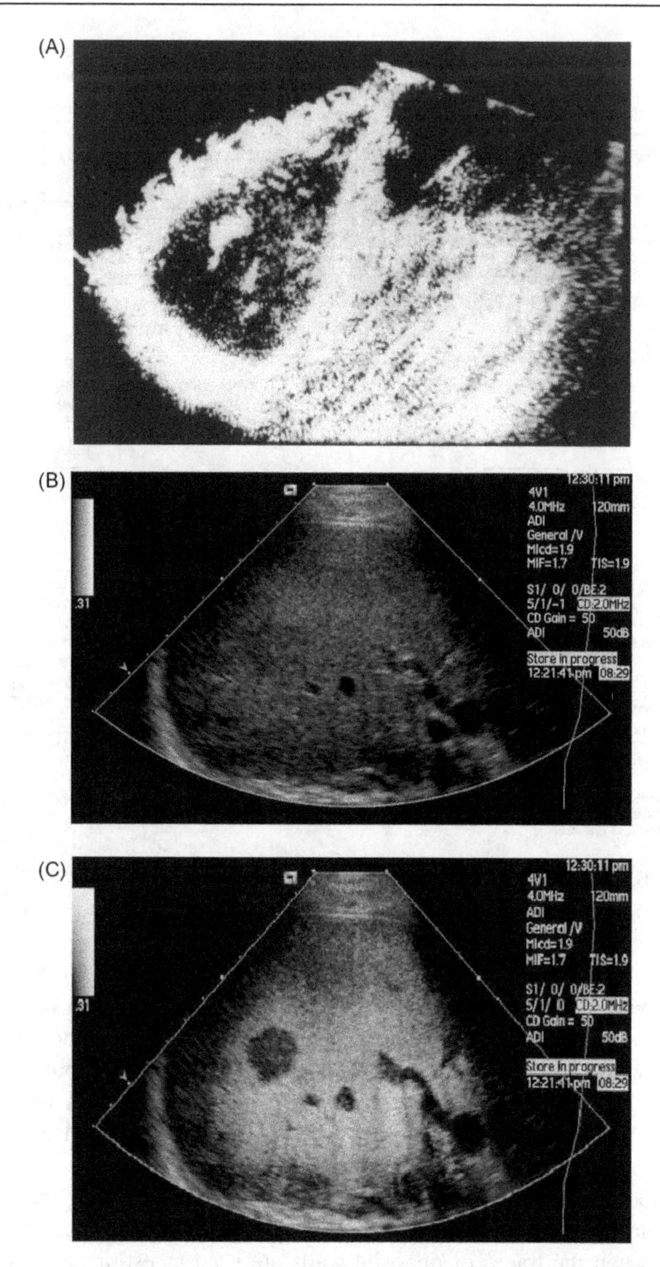

Figure 7.12 Ultrasonic liver scans, illustrating the improvement in quality which has occurred over the last 25 years. The images are (A) a scan which was considered to be of high quality in the early 1970s and which would have been interpreted as supporting the diagnosis of multiple metastases; (B) a scan made with a modern system, in which a metastasis can just be perceived toward the right side of the patient (i.e., toward the left of the image); and (C) a scan of the same patient, in which this lesion is clearly apparent following the administration of an ultrasonic contrast agent [43].

therefore, the Hough transform (HT) may be used to isolate the vessel wall. The HT has been used extensively to characterize straight-line segments and can be extended to generalized features, such as circles and ellipses, for industrial applications. It converts the complex global detection problem in image space into simple local-peak detection in parametric space. Automatic segmentation of the arterial wall offers the additional advantage that the user is not required to outline the boundary of the lumen, often a difficult task in ultrasound images of diseased carotid arteries.

Historically, the earliest ultrasonic studies of physiological motion were made using the pulse-echo technique in M-mode echocardiography. Subsequently, emphasis has largely shifted to the Doppler effect, although direct measurement of target or ensemble movement is also used, with the increase in system sensitivity, partly because it avoids the aliasing limitation inherent in the pulsed Doppler approach.

In principle, the direct measurement of the velocity profile of blood flowing in a vessel provides a more accurate basis for the estimation of the blood flow volume rate than does a single measurement of, say, the maximum blood flow velocity. There have been two approaches to the measurement of the velocity profile. First, a multigated (or infinite gate) pulsed Doppler system has been used. Second, the motion of the speckle, observed by pulse-echo ultrasound to arise from the blood, is tracked over a sequence of consecutive pulse-echo wave-trains. (It is interesting to note that it is the motion of this same speckle that actually enables the pulsed Doppler approach to result in a usable signal.) Having obtained the flow velocity profile and its variation over time (e.g., as the result of cardiac pulsation), the blood flow volume rate may be calculated by assuming the profile to be circularly symmetrical.

Color-flow imaging, which was introduced in the 1980s, similarly may be based on frequency or phase domain (i.e., Doppler) processing or on target motion tracking approaches.

The early Doppler processors were simple auto-correlators, with displays color coded either according to the mean flow velocity or to the amplitude (or power) of the Doppler-shifted signals. More recently, improved velocity estimators, such as the maximum likelihood estimator, have been introduced: they can handle a larger velocity search range which extends to lower values of velocity and with smaller errors.

For the study of blood flow, it is the relatively low amplitude signals from the blood itself which are relevant. The magnitude of these signals can be increased by means of contrast agents and the echoes from solid tissues can be suppressed by second harmonic imaging. When the greatest possible sensitivity is required (e.g., when imaging the microvasculature), the power of the Doppler signals may be used to control the brightness of the image, in distinction from color coding the image according to the velocity of the flow.

In the study of blood flow, the accompanying motion of neighboring solid tissues is a source of difficulty because it gives rise to relatively large amplitude Doppler signals. Various canceling schemes are used, more or less effectively to suppress these echoes. It is usually also necessary to use a high-pass filter

Table 7.1 CT Scan versus Ultrasound: Comparison Chart

Imaging Technique	CT Scan	Ultrasound

Cost	CT scan costs range from $1200 to $3200	Ultrasound procedures cost $100–$1000
Time taken for complete scan	Usually completed within 5 min	Ultrasound usually takes about 10–15 min
Radiation exposure	The effective radiation dose from CT ranges from 2 to 10 mSv, which is about the same as the average person receives from background radiation in 3–5 years.	No radiation
Ability to change the imaging plane without moving the patient	With capability of MDCT, isotropic imaging is possible. After helical scan with Multiplanar Reformation function, an operator can construct any plane	Present
Details of bony structures	Provides good details about bony structures	Ultrasounds are usually not used for bony structures. Instead they are used for internal organs of the body
Details of soft tissues	A major advantage of CT is that it is able to image bone, soft tissue and blood vessels all at the same time	Detailed with advanced technology
Principle used for imaging	Uses X-rays for imaging	High-frequency sound waves (ultrasound) are used for imaging

to remove remaining solid tissue echoes which, fortunately, move more slowly than most of the blood and so give rise to Doppler signals of lower frequencies.

Quite a long time after the introduction of color-flow imaging for the study of blood flow, it was realized that color-coded images of moving solid tissue could also be of clinical utility.

For example, some cardiac lesions are associated with abnormal motion of the myocardium. It turns out to be relatively easy to produce such images, simply by eliminating the solid tissue echo canceler and reducing the cut-off frequency of the pass band filter. The process is known as tissue Doppler imaging.

Finally, a comparison of CT and ultrasound imaging is given in Table 7.1.

References

[1] W.S. Rayleigh, The Theory of Sound, Dover Publications, New York, NY, 1945.
[2] J. Curie, P. Curie, Bulletin de la Société Minéralogique de France 3 (1880) 90 and also published in Comp. Rend. 91 (1880) 294 and 383.
[3] P. Langevin, French Patent No. 505,703, filed September 17, 1917, issued August 5, 1920.
[4] W.D. O'Brien, Jpn. J. Appl. Phys. 37 (1998) 2781.
[5] L.F. Richardson, British Patent No. 9423, filed April 20, 1912, issued March 6, 1913.
[6] L.F. Richardson, British Patent No. 11,125, filed May 10, 1912, issued March 27, 1913.
[7] R.A. Fessenden, Reported in Hunt7 that "this echo-ranging "first" on April 27.
[8] 1914 is recorded in an official report by Captain J. H. Quinan appearing in the U.S. Hydrographic Office Bulletin for May 13, 1914.
[9] R.W. Wood, A.L. Loomis, Philos. Mag. 4 (VII) (1927) 417.
[10] E.N. Harvey, Biol. Bull. 59 (1930) 306.
[11] H. Freundlich, K. Söllner, F. Rogowski, Klin. Wochenschr. 11 (1932) 1512.
[12] R.W. Boyle, J.F. Lehmann, Phys. Rev. 27 (II) (1926) 518.
[13] R.W. Boyle, Nature, London 120 (1927) 476.
[14] R.W. Boyle, G.B. Taylor, Phys. Rev. 27 (II) (1929) 518.
[15] F.O. Schmitt, B. Uhlemeyer, Proc. Soc. Exp. Biol. Med. 27 (1930) 626.
[16] F.W. Kremkau, J. Clin. Ultrasound 7 (1979) 287.
[17] C.R. Hill, Br. J. Radiol. 46 (1973) 899.
[18] J.G. Lynn, R.L. Zwemer, A.J. Chick, A.F. Miller, J. Gen. Physiol. 26 (1942) 179.
[19] J.G. Lynn, T.J. Putman, Am. J. Pathol. 20 (1944) 637.
[20] W.J. Fry, J.W. Barnard, F.J. Fry, R.F. Krumins, J.F. Brennan, Science 122 (1955) 517.
[21] G.R. ter Haar, in: C.R. Hill (Ed.), Physical Principles of Medical Ultrasound, vol. 436, Wiley, New York, NY, 1986.
[22] S. Sokolov, Phys. Z. 36 (142) (1935) and Techn. Physics USSR 2, 522 (1935).
[23] F.A. Firestone, US Patent 2, 280, 226, April 1942.
[24] F.A. Firestone, J. Acoust. Soc. Am. 17 (1946) 287.
[25] F.A. Firestone, J. Acoust. Soc. Am. 18 (1946) 200.
[26] K.T. Dussik, F. Dussik, L. Wyt, Wien. Med. Wschr. 97 (1947) 425.
[27] P.N.T. Wells, in: M. de Vlieger, et al. (Eds.), Handbook of Clinical Ultasound, Wiley, New York, NY, 1978.
[28] W.R. Hendee, J.H. Holmes, in: G.D. Fullerton, J.A. Zagzebski (Eds.), Medical Physics of CT and Ultrasound: Tissue Imaging and Characterization, American Institute of Physics, New York, NY, 1980, p. 298.
[29] J.J. Wild, Surgery 27 (1950) 183.
[30] L.A. French, J.J. Wild, D. Neal, Cancer 3 (1950) 705.

[31] J.J. Wild, J.M. Reid, J. Acoust. Soc. Am. 25 (1953) 270.
[32] G.D. Ludwig, F.W. Struthers, Electronics 23 (1950) 172.
[33] D.H. Howry, W.R. Bliss, J. Lab. Clin. Med. 40 (1952) 579.
[34] J.J. Wild, J.M. Reid, Science 115 (1952) 226.
[35] J.J. Wild, J.M. Reid, Am. J. Pathol. 28 (1952) 839.
[36] W.P. Mason, Phys. Rev. 72 (1947) 854.
[37] S. Roberts, Phys. Rev. 25 (1947) 890.
[38] B. Jaffe, R.S. Roth, S. Marzullo, J. Appl. Phys. 25 (1954) 809.
[39] J.M. Reid, J.J. Wild, IRE Trans. Ultrason. Engr. PGUE-5 (1957).
[40] I. Donald, T.G. Brown, Br. J. Radiol. 34 (1961) 539.
[41] I. Donald, Ultrasound Med. Biol. 1 (1974) 109.
[42] I. Donald, Med. Biol. Illustr. 14 (1964) 216.
[43] P.N.T. Wells, Phys. Med. Biol. 51 (2006) R83.
[44] S. Golemati, J. Stoitsis, T. Balkizas, K.S. Nikita, Engineering in Medicine and Biology 27th Annual Conference Shangai, China, September 1–4, 2006.

8 Raman Spectroscopy in Medicine

8.1 Introduction

Successive advances in biomedical technology have been driven by the need to provide objective and quantitative diagnostic information. Novel biomedical applications of optical spectroscopy, such as fluorescence, reflectance, and Raman scattering, can provide information about the composition of tissue at the molecular level. Among these techniques, Raman spectroscopy can provide the most detailed information about the chemical composition of the tissue under study. The progression of disease is accompanied by certain chemical change. In this context, Raman spectroscopy can provide the physician with valuable information for diagnosing disease. Since light can be delivered and collected rapidly via optical fibers (these can be incorporated into catheters, endoscopes, cannulas, and needles, to say about few), Raman spectroscopy can be performed *in vivo* in real time.

Diagnostic applications of Raman spectroscopy currently under investigation are widespread. For example, Raman spectroscopy may be used to monitor blood analyzes noninvasively. Further, it may be used to perform minimally invasive, real-time, tissue diagnosis *in vivo*, in cases where biopsy cannot be performed readily, such as coronary artery disease and Alzheimer's disease, or where a high incidence of false positive screening tests leads to unnecessary biopsy procedures, as in a case of breast cancer. The purpose of this chapter is to discuss the potential of Raman spectroscopy to provide real time, objective, and at the same time quantitative diagnostic information *in vivo*.

There is no doubt that the early detection of cancers, monitoring of the effect of various agents on the skin, determination of atherosclerotic plaque composition, and rapid identification of pathogenic microorganisms is vitally important.

Taken from the viewpoint of clinicians and medical analysts, the potential of Raman spectroscopic techniques as new tools for biomedical applications is discussed here and a path for the clinical implementation of these techniques is proposed.

8.2 From the History of Raman Effect

Raman spectroscopy is named after the famous Indian physicist Sir Chandrasekhara Venkata Raman (Figure 8.1). Along with K.S. Krishnan, he found in 1928 that when

Medical Imaging Technology. DOI: http://dx.doi.org/10.1016/B978-0-12-417021-6.00008-3

Figure 8.1 Professor Sir C V Raman.

a beam of light transverses a transparent medium, a small fraction of that beam will emerge from the compound at right angles to and of a different wavelength from the original beam [1].

Although he was a brilliant student, higher education was considered below his caste. So in 1907 he went to Calcutta to join the Indian Financial Civil Service as Assistant Accountant General. Raman, a "self-made scientist," with intense drive, worked early mornings and evenings in the underutilized facilities of the Indian Association for the Cultivation of Science, in Calcutta, to study problems in acoustics. He conducted research independently, for nearly 10 years, and established his reputation. In 1917, Raman was appointed professor at the University of Calcutta.

Somewhat latter, in 1921, on his first voyage outside India, to Oxford, he performed some experiments and published a note entitled "The color of the sea" in *Nature*. He showed that the color of the ocean is independent of sky reflection or absorption and is instead due to scattering. It is a beginning of his scientific research into the scattering of light. The work led to his discovery of the Raman effect in 1928, and he was awarded the Nobel prize just 2 years later (a record time). He was the first Asian, and the only Indian, to receive this award.

Among scientists, Raman is best known for his strikingly ingenious yet simple experimental design, the lucidity of his ideas, and the depth of his observations.

The Raman effect is a fundamental process in which energy is exchanged between light and matter. When light impinges on a substance it can be scattered or absorbed. Most of the scattered light will have the same frequency as that of the incident light. However, a small fraction of the incident light can go into setting molecules in the material into vibration. The energy for this must come from the incident light. Since light energy is proportional to frequency, the frequency change of this scattered light must equal the vibrational frequency of the scattering molecules. This process—energy exchange between scattering molecules and incident light—is known as the Raman effect.

8.3 The Principle of Raman Scattering

The process of Raman scattering, from the energy level point of view, can be considered as the transition of a molecule from its ground state to an excited vibrational state, accompanied by the simultaneous absorption of an incident photon and emission of a Raman scattered photon (Figure 8.2) [2].

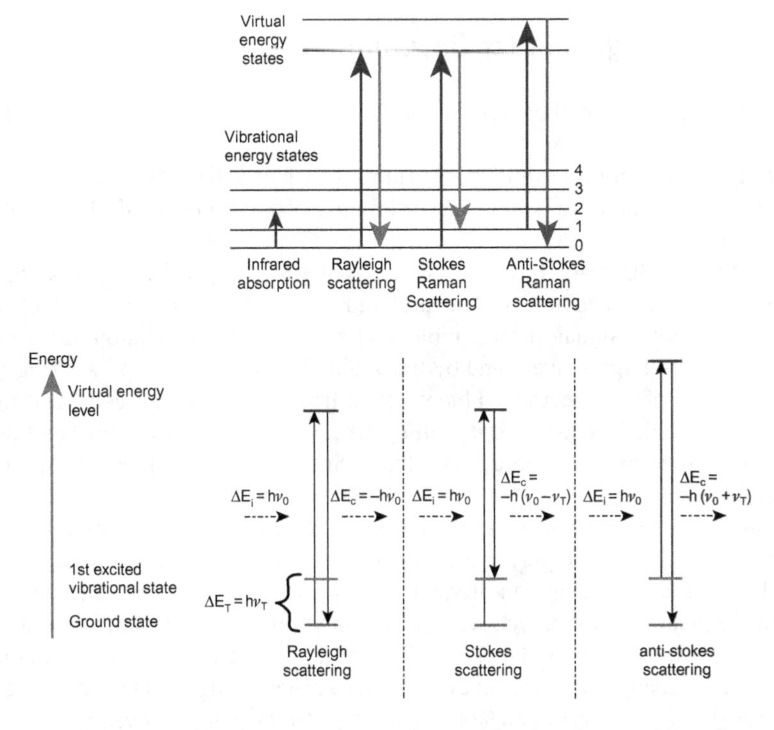

Figure 8.2 The process of Raman scattering can be considered as the *transition of a molecule from its ground state to an excited vibrational state*. This transition is accompanied by the simultaneous absorption of an incident photon and emission of a Raman scattered photon. The horizontal lines indicate vibrational energy levels. Ground electronic state and excited electronic states are shown. The diagrams show how a molecule in the ground state can make a transition from the lowest vibrational level to the first excited vibrational level by means of Raman scattering. Black up-arrows indicate the frequency of the laser excitation light; black down-arrows indicate frequency of the Raman scattered light. The difference in length between black up- and down-arrows indicates molecular vibration frequency. It should be noted here that Raman scattering at different excitation wavelengths, UV, visible, and near-IR, produces the same change in vibrational energy. Therefore, the excitation wavelength can be chosen to avoid spectral interference by fluorescence. For visible excitation, the fluorescence light frequency and Raman scattered light frequency are similar. This leads to intense fluorescence background in visible excitation Raman spectra. Near-IR light has too low a frequency to excite fluorescence, while for UV excitation, the fluorescence light frequency is much lower than the Raman scattered light frequency. Hence, fluorescence background in the Raman spectrum can be reduced by using UV or near-IR excitation.

The Raman scattered light can be collected by a spectrometer and displayed as a "spectrum", i.e., intensity versus frequency change. Since each molecular species has its own unique set of molecular vibrations, the Raman spectrum of a particular species will consist of a series of peaks or "bands," each shifted by one of the vibrational frequencies characteristic of that molecule.

8.4 Technology and Its Development

The technology and methodology to acquire and interpret the necessary Raman spectra needs to be developed.

During a long period of time, tissue fluorescence and the lack of appropriate and sensitive equipment were serious obstacles in the development of clinical Raman applications.

Over the last 30 years, however, the field of Raman spectroscopy has seen tremendous impressive advances. The problem of tissue fluorescence, which overwhelms the Raman signal of most biological samples upon excitation in the visible region, has been largely overcome by the availability of instruments working in the near-IR region of the spectrum. Fluorescence from biomolecules, cells, and tissues can also be avoided by laser excitation in the deep UV. It was established that, for the Raman spectrum, a fluorescence-free window exists with excitation below 270 nm [3–6].

As an example, the Raman spectrum of cholesterol, a representative biological molecule, is shown in Figure 8.3. The spectrum is a plot of the scattered light intensity versus its change in frequency, relative to that of the incident light. Raman frequency shifts are conventionally measured in wavenumbers (cm^{-1}), a unit convenient for relating the change in vibrational energy of the scattering molecule to the change in frequency of the scattered light. One cm^{-1} equals 30,000 MHz; *this is typically 10,000 times smaller than the frequency of the light*

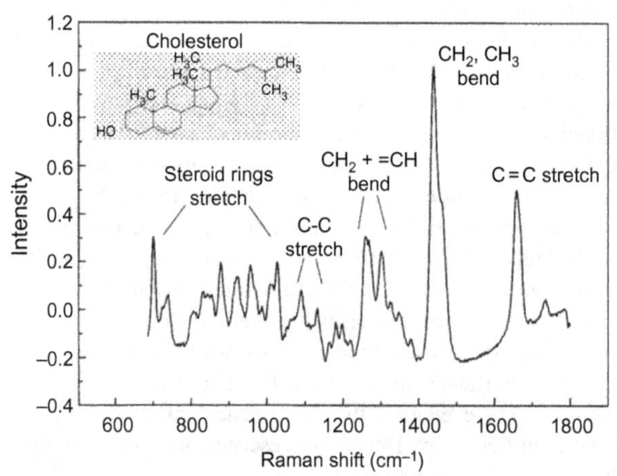

Figure 8.3 Near-IR Raman spectrum of cholesterol molecule. Shown are typical vibrational bands. The background has been removed by subtracting [3,4].

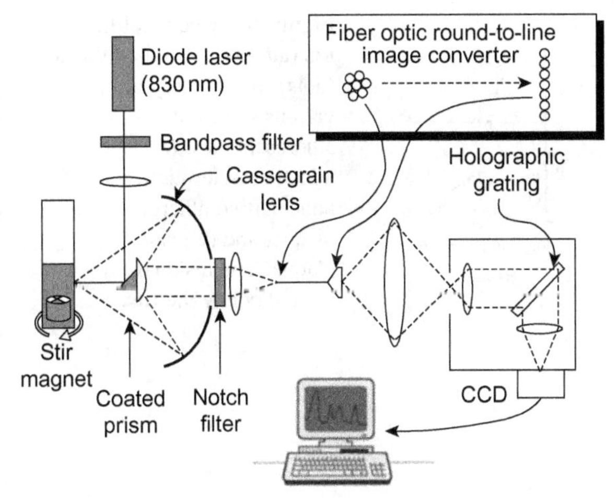

Figure 8.4 Raman spectroscopic system used for the analysis of blood analytes. The reflective objective has high collection efficiency. Inset: The collection efficiency and high throughput are maintained, while preserving resolution, by configuring the f-number matched fiber bundle into a linear array at the spectrograph entrance. This linear array of fibers forms the spectroscopic slit that determines resolution.

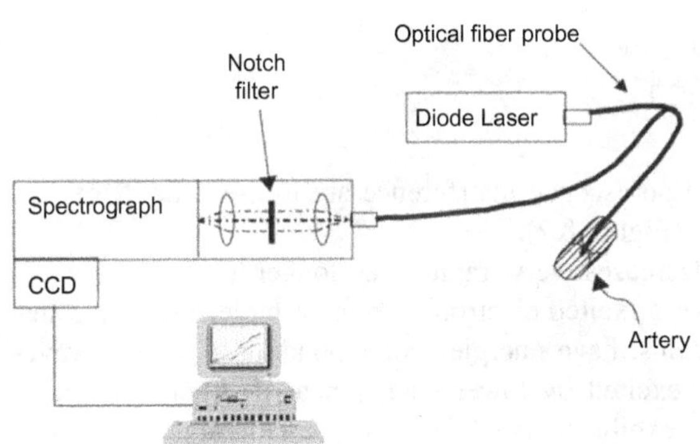

Figure 8.5 Clinical Raman system designed for rapid data acquisition, portability, and safety in a hospital environment.

itself. Each band of scattered light in the cholesterol Raman spectrum is characteristic of molecular vibrational motions, which, taken all together, are unique for cholesterol.

If a sample of biological tissue contains cholesterol, characteristic peaks (e.g., peak at 1440 cm^{-1} is due to the CH_2 and CH_3 deformation vibrations, and the peak at 1670 cm^{-1} is due to C=C stretching vibrations) will be present in its Raman spectrum. In other words, the molecular structure and composition of a material under study is encoded as a set of frequency shifts in the Raman scattered light. Thus, the Raman spectrum can provide a "fingerprint" of a substance from which the molecular composition can be determined.

In Raman spectroscopy, the incident light is often referred to as "excitation" light, because it excites the molecules into vibrational motion. Ultraviolet, visible, and infrared light can each be used for Raman excitation.

When visible excitation is used, tissue exhibits strong, broadband fluorescence that can obscure the tissue Raman spectrum.

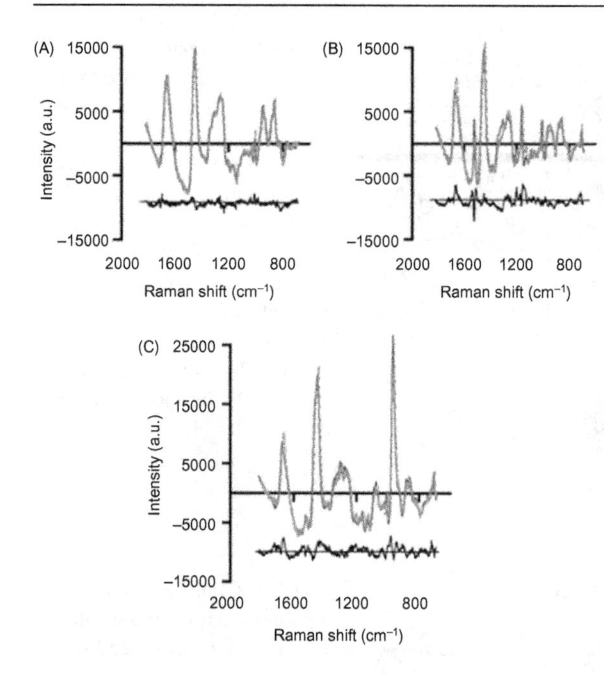

Figure 8.6 Chemical model fits (curve) and spectral data (dots) for coronary arteries in various stages of atherosclerosis: (A) nonatherosclerotic tissue, (B) noncalcified atheromatous plaque, and (C) calcified plaque. The residuals are plotted below the fits.

Two strategies for reducing fluorescence interference are to use near-IR excitation or UV resonance excitation (Figure 8.2).

Importantly, fluorescence decreases very rapidly at longer excitation wavelengths. This is because the lowest excited electronic states of molecules, the states from which fluorescence originates, have energies corresponding to visible wavelengths, and so they cannot be excited by lower energy near-IR light. Therefore, most materials, including tissue, exhibit reduced fluorescence emission as the excitation wavelength increases into the near-IR region. Thus, fluorescence interference in tissue Raman spectra can be greatly reduced by using near-IR excitation.

An alternate strategy is to use excitation wavelengths in the UV. It was established that background fluorescence is essentially suppressed in tissue Raman spectra for excitation wavelengths below about 270 nm. An additional advantage of UV excitation light is that it can resonantly excite the electronic states of the molecule. In this case, the Raman scattering for certain vibrations is enormously enhanced, providing larger signals. Disadvantages of UV resonance excitation are that the tissue penetration depth is of the order of microns, so this technique is only applicable for probing superficial abnormalities, and that UV light at these wavelengths can induce mutagenesis [5–7].

In general, Raman spectra of tissue consist of relatively narrow bands, typically 10–20 cm^{-1} in width, which exhibit the presence of many biochemicals. The relative contributions of these biochemicals to the whole Raman spectrum are proportional to their relative abundance in the tissue. This is the basis for the quantitative nature of the information Raman spectroscopy can provide for diagnosis. The quantitative nature of Raman spectra, combined with the ability to provide unique "signs"

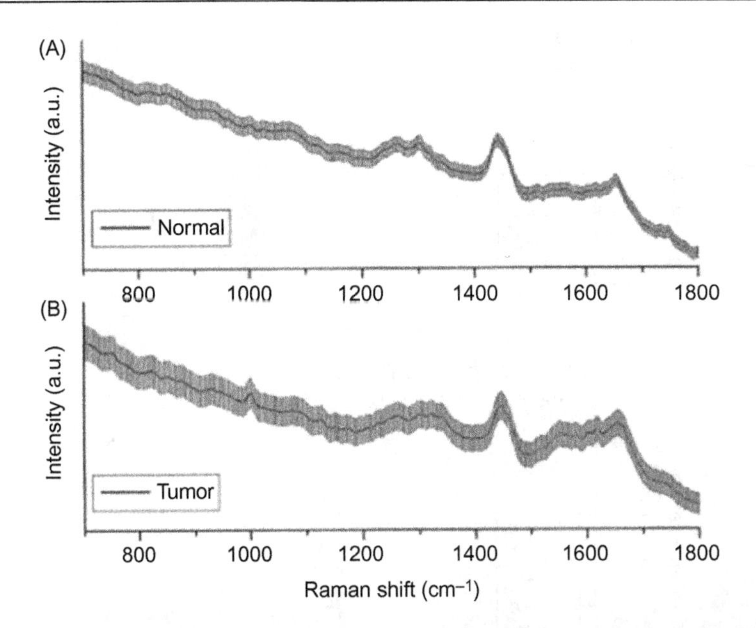

Figure 8.7 Mean Raman spectra (solid line) 1 SEM (gray area) were obtained from single representative samples of normal tissue (A) and tumor (B) (SCC) by performing multiple measurements ($n = 10$) for each sample at different tissue orientations (epithelial and stromal sides of the tissue). All spectra were acquired in 5 s with 785 nm excitation and corrected for the spectral response of the system.

of the biochemicals present in tissue, demonstrate the potential of Raman spectroscopy to provide objective, quantitative diagnostic information for tissue analysis.

Raman spectroscopy had been widely used for chemical and molecular analysis for many years. At the same time, its application to biomedical problems is relatively recent. Raman studies of biological tissue have been facilitated over the past 20 years by advanced technology, particularly in the areas of lasers and detectors. However, the true success of these studies in providing insight into the molecular basis of disease and showing the potential for medical diagnostic applications has relied on advanced methods of analyzing tissue Raman spectra. In order to exploit the full potential of Raman spectroscopy, methods of analysis that employ the full information content of the Raman spectrum (not the prominent peaks only) must be used. The problem of extracting the full, diagnostically useful information that tissue Raman spectra contain, can be divided on three distinct levels: statistical, chemical, and morphological:

a. *Statistical analysis*: Up to now, much of the work in this field has relied largely on empirical methods of analysis, such as correlating tissue type with prominent spectral bands. However, mathematical methods such as principal component analysis can be used to characterize the full range of spectral variations. The principal components can then be fitted to the observed Raman spectral contours. Correlating the fit coefficients

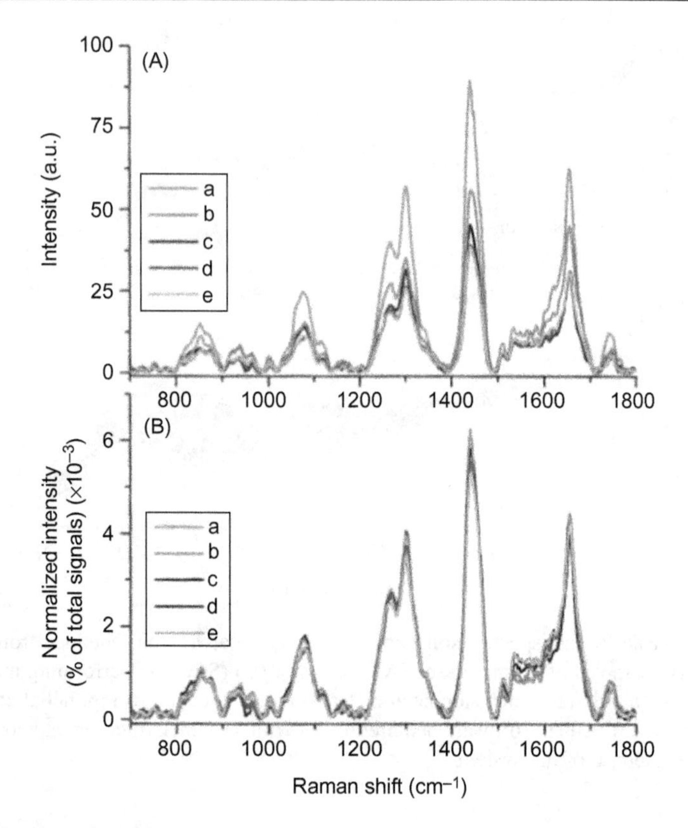

Figure 8.8 Effect of a specimen size on Raman spectra for normal bronchial tissue. (A) Before normalization for sample sizes: (a) $8 \times 8 \times 1.5\ mm^3$; (b) $5 \times 5 \times 1.5\ mm^3$; (c) $4 \times 4 \times 1.5\ mm^3$; (d) $3 \times 3 \times 1.5\ mm^3$; (e) $1 \times 1 \times 1.5\ mm^3$. The larger specimen has a stronger Raman signal. (B) After normalization, all Raman spectra for different specimen sizes display almost identical patterns with intensity variations of 10–20% for major Raman peaks.

with tissue type can be used to classify tissue by diagnostic category. Additionally, features of the principal components can be used as a guide in identifying key biochemical or morphological constituents of the tissue or disease under study.

b. *Chemical analysis*: Raman spectroscopy can provide information about the chemical composition of tissue. A Raman spectrum can be modeled as a superposition of the spectra of its chemical constituents, thereby providing quantitative chemical information. Chemical information is of importance: the onset of disease is accompanied by biochemical changes, and Raman spectroscopy can be an excellent method for detecting biochemical changes (even if they are vanishingly small) that may occur in either the cellular or the extracellular compartments of tissue. At present, few techniques are capable of providing detailed *in vivo* biochemical analysis of tissues. Such techniques have an enormous potential, both in diagnosing disease and in understanding its origins and evolution.

c. *Morphological analysis*: Quantitative methods to extract information regarding microscopic constituents present in a tissue sample, i.e., the morphological structures, are also

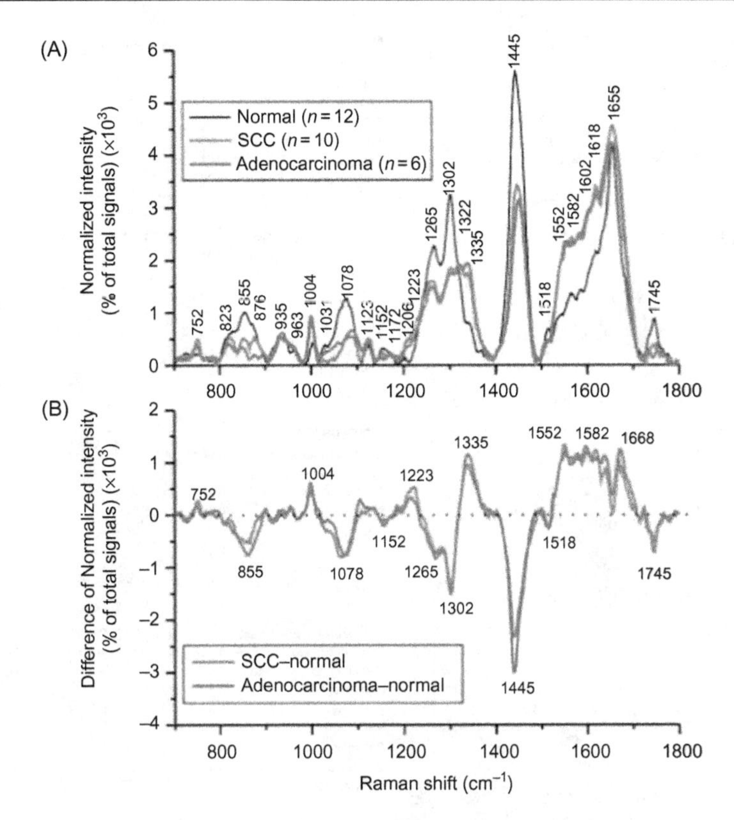

Figure 8.9 (A) The mean Raman spectra of normal bronchial tissue ($n = 12$) and malignant adenocarcinoma ($n = 6$) and SCC ($n = 10$) bronchial tissue samples. Each spectrum was normalized to the integrated area under the curve to correct for variations in absolute spectral intensity. (B) Difference spectra were calculated from the mean spectra: SCC minus normal and adenocarcinoma minus normal.

possible. Pathologists make a tissue diagnosis by assessing the presence, absence, or relative abundance of such constituents, which include various types of cells, fibers, and mineral deposits. Raman spectroscopy has the potential to provide such morphological information without tissue removal, in real time and in a quantitative and objective manner. Further development of this interesting area is needed.

8.5 Clinical Applications of Raman Spectroscopy

8.5.1 Clinical Raman System

Laboratory Raman system for blood analyses (Figure 8.4) has been designed to study how best to optimize delivery and collection geometries in liquid samples of varying turbidity—whole blood, blood serum, interstitial fluids, and saline solution

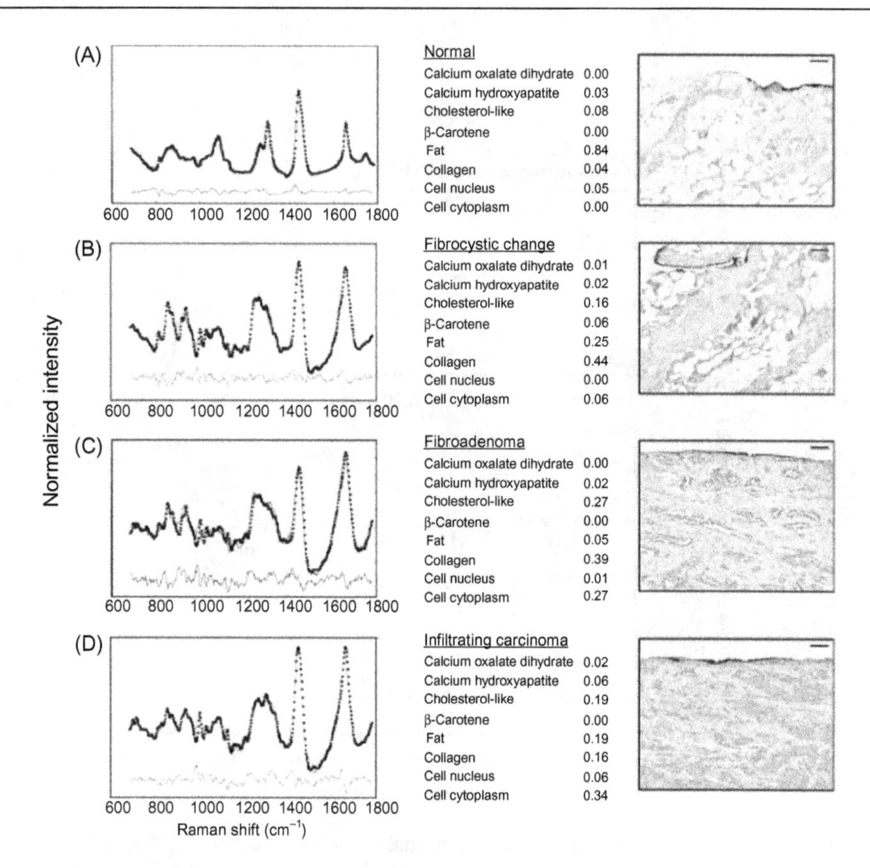

Figure 8.10 Raman spectra: (A) normal breast tissue, (B) fibrocystic change, (C) fibroadenoma, and (D) infiltrating carcinoma.

[3]. For this, the delivery and collection geometries can be precisely varied. When used with multivariate techniques, physiological concentrations of analyses (e.g., glucose) at sub-millimolar levels can be measured (Figure 8.5). Noting that water is 55 M, this represents a Raman concentration measurement sensitivity of 10 ppm, illustrating the potential sensitivity of Raman spectroscopy for precise, quantitative measurements. The system uses a tunable diode laser operating at 830 nm and 500 mW power, a holographic imaging spectrograph, liquid nitrogen-cooled deep depletion CCD detector, and special collection optics.

8.6 Atherosclerotic Plaques, Breast Cancer, Lung Cancer, and Skin Cancer

The characterization of atherosclerotic plaques is an example of an application in which the feasibility of Raman spectroscopy to obtain certain information about

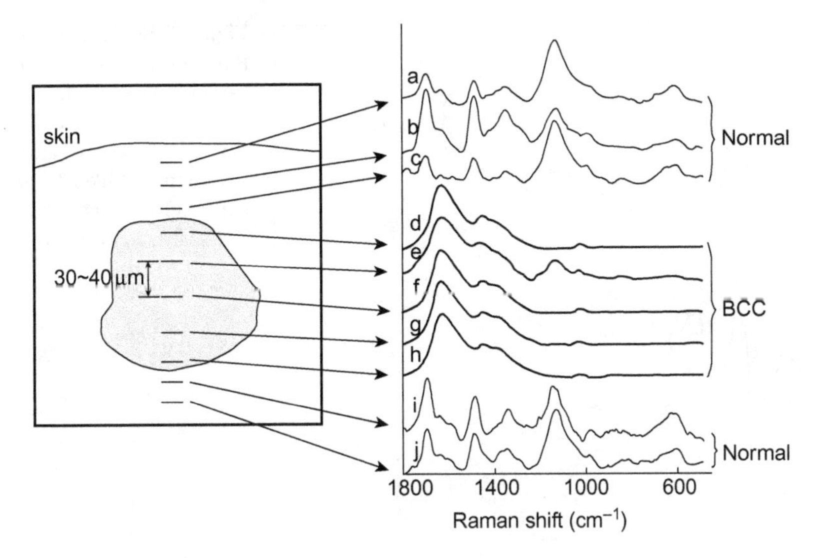

Figure 8.11 Confocal Raman profiles of skin tissue with an interval of 30–40 μm.

tissue composition has been demonstrated [3]. Information extracted from Raman data essentially improves correct clinical decision.

A detailed *ex vivo* analysis of the molecular composition of a plaque can be obtained by Raman spectroscopy [6]. For example, it is not certain whether the chemical composition of a plaque has an effect on the occurrence of re-stenosis after interventions such as (balloon) angioplasty or stenting and if a particular drug regime can be particularized to find and promote therapeutic efficiency. Hence, knowledge of the plaque chemical composition might play a role in the choice of a certain therapy.

Further, in recent years much of the debate about the occurrence of heart attacks has revolved around the questions of whether all plaques are putting a patient at risk and whether a distinction should be made between different types of plaques. For example, "vulnerable" plaques (composed of lipid pools separated from the bloodstream by a fibrous cap) are responsible for most, often fatal, heart attacks.

The rupturing of the fibrous cap exposes the blood to the highly thrombogenic contents of the lipid pool, resulting in a blood clot that occludes the artery. At the other end of the spectrum are the "stable" plaques that may be highly calcified and often occlude a significant percentage of the blood. These latter plaques cause fatal heart attacks less frequently because they give distinct signs in the form of symptoms such as chest pain during physical exercise and they do not cause formation of blood clots that can suddenly occlude a blood vessel.

Figures 8.6–8.11 illustrate the potential of Raman spectroscopy in diagnosis of cancer in breast, lung, and skin tissues.

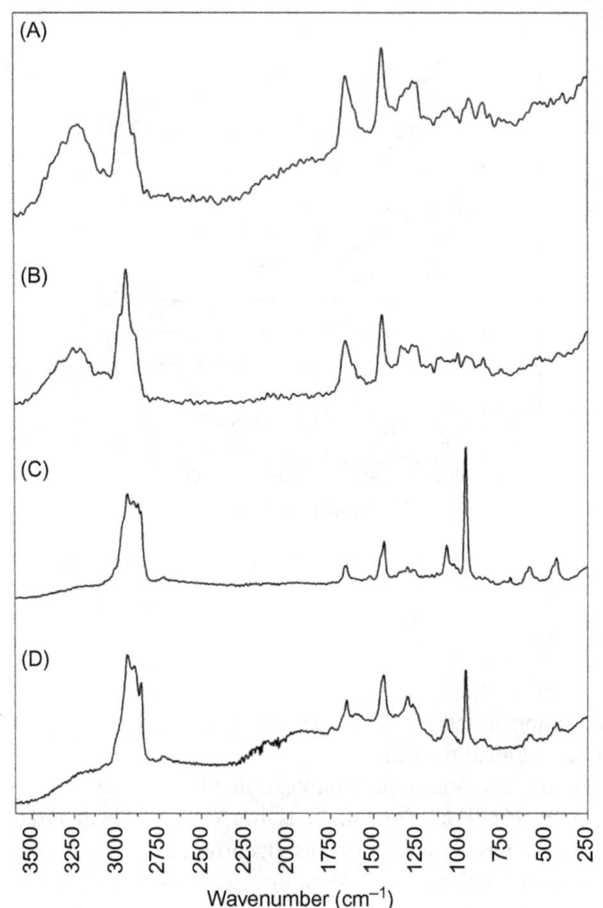

Figure 8.12 Representative Raman spectra of human tissues: (A) arm skin, (B) an intact artery, (C) calcified plaque, (D) bone (femur). (FT Raman Bruker RFS/100, Nd:YAG:1064 nm, 450 mW, 128 scans, 4 cm^{-1}) [9].

8.7 Blood Vessels

Atherosclerosis is one of the most common causes of death in developed and developing countries. Its etiology is still not fully known. Generally, the atherosclerotic process is characterized by cholesterol aggregation in the middle layer of a vessel. In the later stage of disease, wall calcification is observed [8,9]. This results in vessel stenosis and reduced blood flow. The atherosclerotic processes are mentioned as one of aortic aneurysm causative factors. Moreover, these factors can also lead to arterial thrombosis.

Raman spectroscopy is used for assessing arterial lesions resulting from pathological processes and provides detailed information about, for example, the composition of an athermanous plaque.

The Raman spectrum of healthy aorta is dominated by vibrational bands of its structural proteins—collagen and elastin (Figure 8.12) whose compositions change with disease progression, whereas the Raman spectrum of atherosclerotic tissue is dominated by the lipid bands. Increased intensities of the bands corresponding to cholesterol and its esters (880, 850, 720, and 700 cm^{-1}) are observed. The higher lipid contents are also confirmed by clearly broader bands of the CH_2 and CH_3 stretching and deformation vibrations as well as shifts of their maxima to lower wavenumbers.

Moreover, in the spectra of a calcified atheromatous plaque (Figure 8.12C), intense bands of the phosphate group vibrations are present, contrary to the spectra of healthy vessel walls.

The main calcium compound that occurs in its deposits in blood vessel walls is carbonate apatite of low crystallinity similar to that of bone or dentin mineral.

Raman scattering with 488 nm excitation light can be employed to study blood plasma samples from several patients and observed differences in the carotenoid spectral regions between patients with cancer and normal patients.

References

[1] C.V. Raman, K.S. Krishnan, Nature 121 (501) (1928) 3048.
[2] Int. Rev. Phys. Chem. 7 (1988).
[3] E.B. Hanlon, R. Manoharan, T.-W. Koo, et al., Phys. Med. Biol. 45 (2000) R1.
[4] Z. Huang, et al., Int. J. Cancer 107 (2003) 1047.
[5] A. Downs, Sensors 10 (2010) 1871.
[6] L.P. Choo-Smith, et al., Biopolym. (Biospectroscopy) 67 (2002) 1.
[7] P.J. Lambert, A.G. Whitman, O.F. Dyson, S.M. Akula, Virol. J. 3 (51) (2006) 1.
[8] A. Mohadevan-Jansen, R. Richards-Kortum, Proceedingsof the 19th International Conference—IEEE/EMBS, October 30, 1997, Chicago, IL.
[9] S. Olsztyńska-Janus, M. Gąsior-Głogowska, K. Szymborska-Małek, et al., Acta Bioeng. Biomech. 14 (2012) 121.

CPSIA information can be obtained
at www.ICGtesting.com
Printed in the USA
LVHW080442211222
735623LV00004B/207